A Brief History of I

MW00773681

Pharmacy has become an integral part of our lives. Nearly half of all 300 million Americans take at least one prescription drug daily, accounting for $250 billion per year in sales in the US alone. And this number doesn't even include the over-the-counter medications or health aids that are taken. How did this practice become such an essential part of our lives and our health?

A Brief History of Pharmacy: Humanity's Search for Wellness aims to answer that question. As this short overview of the practice shows, the search for well-being through the ingestion or application of natural products and artificially derived compounds is as old as humanity itself. From the Mesopotamians to the corner drug store, Bob Zebroski describes how treatments were sought, highlights some of the main victories of each time period, and shows how we came to be people who rely on drugs to feel better, to live longer, and look younger. This accessible survey of pharmaceutical history is essential reading for all students of pharmacy.

Bob Zebroski is Professor of History and Chairperson of the Liberal Arts Department at the St. Louis College of Pharmacy, one of the largest and oldest schools of pharmacy in the US, which just celebrated its sesquicentennial anniversary.

A Brief History of Pharmacy
Humanity's Search for Wellness

Bob Zebroski

Routledge
Taylor & Francis Group

NEW YORK AND LONDON

First published 2016
by Routledge
711 Third Avenue, New York, NY 10017

and by Routledge
2 Park Square, Milton Park, Abingdon, Oxon OX14 4RN

Routledge is an imprint of the Taylor & Francis Group, an informa business

© 2016 Taylor & Francis

The right of Bob Zebroski to be identified as author of this work has been asserted by him in accordance with sections 77 and 78 of the Copyright, Designs and Patents Act 1988.

Library of Congress Cataloging in Publication Data
Zebroski, Bob.
A brief history of pharmacy: humanity's search for wellness / Bob Zebroski.
pages cm
Includes index.
1. Pharmacy—History. I. Title.
RS61.Z47 2015
615.1—dc23
2015009903

ISBN: 978-0-415-53783-4 (hbk)
ISBN: 978-0-415-53784-1 (pbk)
ISBN: 978-1-315-68583-0 (ebk)

Typeset in Bembo
by Swales & Willis Ltd, Exeter, Devon, UK

Printed and bound in the United States of America by Publishers Graphics, LLC on sustainably sourced paper.

"Time and history tells and shows all—where we have been and where we might go. It provides insight to ourselves, other people and the world in which we live. For us it can provide the opportunity to judge our actions and motivations and perhaps give us a glance into the future. Retrospect for prophecy. It is crucial in giving us a sense of self."

John Klemmer

Contents

Figures

Tables

Introduction
The Essentials of Pharmacy

When I began teaching a History of Medicine and Pharmacy course in 1998 a friend and colleague passed the following electronic message to me entitled, "The History of Medicine," which in its own way outlines the story of the book you are about to read.

The History of Medicine

2000 BC	Here, eat this root.
1000 AD	That root is heathen. Here, say this prayer.
1850 AD	That prayer is superstition. Here, drink this potion.
1940 AD	That potion is snake oil. Here, swallow this pill.
1985 AD	That pill is ineffective. Here, take this antibiotic.
2000 AD	That antibiotic doesn't work anymore. Here, eat this root.

While comical, the e-mail contains a grain of truth that led me to write this book. This book is for people who want a brief introduction to the history of pharmacy. Its origin came about from my students' desire to have an accessible account about the history of pharmacy that deals with it from its origins to the present. This book is meant to be a starting point, a synthesis of the best scholarship to introduce its readers to the history of pharmacy and will provide a bibliography (in the form of chapter endnotes) for further study of any desired topic. As we shall see, the material conditions of the cultures and civilizations discussed in the book vary over time and place, but the fundamental human quest for health and well-being remains universal. This book is a humble attempt to tell this fascinating story.

The Essentials of Pharmacy

What is pharmacy? The great American historian of pharmacy, George A. Bender, defined it as follows: "Pharmacy, the profession of the art and science of preparing, preserving, compounding, and dispensing medicines, indeed has a proud heritage—an unequalled record of service to humanity almost as old as the human race itself."[1] The origin of the word "pharmacy" can be traced back to the ancient Egyptian word *ph-ar-maki*, which meant granter of security. It is more likely taken from the Greek word *pharmakon* which, ironically, means either remedy or poison.[2] The essential people are pharmacists, who over time, were variously known as shamans, priests, healers, diviners, physicians, apothecaries, pharmaceutists, chemists, and druggists; in other words people who prepared and administered drugs to patients. In fact, in the United States, apothecaries did not get the name "pharmacist" until an officer of the then American Pharmaceutical Association, Edward Parrish, proposed it and the term was officially adopted by the organization in 1867.[3] It is important to note that from prehistoric times until at least the Middle Ages, our ancestors entrusted their health to "healers." These healers performed all of the functions that physicians, pharmacists, nurses, and other health care specialists perform today.

The Eastern tradition of ancient healing explored the sacred and essential relationship between the pharmacist and the patient further. The legendary sage of ancient Indian pharmacy, Caraka (circa first century CE), in his work the *Caraka-Samhita* wrote, "the physician, drug, nurse (or attendant), and patient are the four pillars upon a qualified synthesis of which is supported the treatment of diseases."[4] Since humans have made their appearance on this planet, the essential components of pharmacy, i.e. human beings and drugs (albeit in natural form) have coexisted. Thus, the essence of pharmacy is people and drugs; one without the other is not pharmacy. In our age of high tech medicine it is easy to forget these essentials and we pay a high price for it. As John Naisbitt and Patricia Aburdence have observed, as we move more toward high technology the consumer demands more high touch or personal service, not less.[5] In other words, true healing involves the humanity of the pharmacist and as we progress toward genetic therapies and more personalized medicine, patients will require more humanity from their health care providers. Putting back the "care" in health care will be the key factor in pharmacy's survival as a profession.[6] History runs in cycles; what was once in fashion often returns in similar form at some future point in time, the eternal return.

Cura Personalis or **Treating the Whole Patient**

Hermann Peters, a late nineteenth-century German historian who pioneered the modern study of the history of pharmacy, argued that for our ancestors, "Disease was the 'soul' of one object attacking another."[7] Today, Dr. Rachel Naomi Remen reminds us that:

> At the beginnings of medicine, the shamans, or medicine men, defined illness not in terms of pathology but in terms of the soul. According to these ancients, illness was "soul loss," a loss of direction, purpose, meaning,

mystery, and awe. Healing involved not only the recovery of the body but the recovery of the soul.[8]

According to Dr. Remen, "Perhaps what is needed is not only to learn good medicine but to become good medicine. As a parent. A friend or a doctor. Sometimes just being in someone's presence is strong medicine."[9] Similarly, the late Dr. Lewis Thomas observed, "Hope is itself a kind of medicine."[10] Shamans intuitively understood that they were treating "the whole patient." The Romans later developed a term for caring for the whole patient and called it *cura personalis*. The language that heals the soul is the language and symbolism of what it means to be alive and human. This primeval system of healing the whole person falls under the contemporary classification of the "bio-cultural model" and became the dominant model from prehistoric times until the advent of germ theory in the late nineteenth century. The biomedical model that emerged with germ theory continues to be the dominant paradigm around the world today. Nonetheless, it is important for pharmacists and all of us today to understand that the biomedical model emerged from the bio-cultural model just as science emerged from philosophy. Pharmacy at its best is scientifically based, but humanistic in its treatment of patients. In fact, humanistic treatment of patients is what today's personalized patient-centered care is all about. That is why pharmacists must take a solemn oath to put their patients' welfare before their own as professionals. Norman Cousins observed, "Medical science may change, but the need to understand and deal with human beings remains constant . . . Moral values, an important aspect of human uniqueness, are not perishable or evanescent or replaceable."[11]

Moreover, Dr. Lewis Thomas, the American physician and Pulitzer Prize winning author points to the origins of the words "patient" and "doctor" as especially meaningful.[12] In its most essential form a "patient" is one who suffers; while a "doctor," as in a doctor of pharmacy, is one who teaches. Today doctors of pharmacy are clinicians who render pharmaceutical care and engage in medication therapy management with their patients, which in large part means teaching their patients about how to get the most out of their medications in order to achieve optimal health outcomes. Once again, while material conditions may change over time, the essential act of a pharmacist treating a patient remains a sacred act of mutual trust and goodwill whether it was back in prehistoric times or today.

How to Use This Book

For those readers who want to understand the origins of pharmacy and how it evolved, the first eight chapters trace the common history of healing from the Prehistoric Era through the Late Middle Ages emphasizing pharmacy as a specialized field of medicine developing according to a bio-cultural medical model. During these millennia all of the specialties involved in healing today—the physician, surgeon, physician's assistant, pharmacist, nurse, and other allied health professionals—were vested in the person of the "healer," who more often times than not was also a priest. While some ancient civilizations such as the Mesopotamians, Egyptians, and Romans sometimes made references to specialists who gathered, prepared, stored, and administered drugs, other civilizations did not. During the late eighth century when Islamic civilization experienced a golden age of scientific discovery, the first public pharmacy opened in Baghdad, and was an integral part of a deliberate public health program in which apothecaries and physicians played separate and distinct roles. Still, several centuries passed before this development impacted Europe. Pharmacists, or apothecaries as they were called then, did not gain any official recognition as an autonomous specialty of medicine until the Holy Roman Emperor Frederick II issued his edict called the *Constitutiones* around 1240, which formally recognized apothecaries as distinct from physicians. While this edict influenced the development of pharmacy on the Continent, it did not affect other nations. For example, in Britain, apothecaries did not gain similar autonomy until the early seventeenth century under an edict issued by King James I in 1617. The first part of the book is for those readers who want the full story to experience how the past and present speak to each other.

For those readers who want to focus on the modern era, especially on the development of American pharmacy, Chapters 9–16 are for you. These chapters show how the autonomous apothecary who emerged during the Renaissance became the eighteenth century's chemist, who in turn became the nineteenth century's pharmaceutist and druggist, and who was ultimately destined to become today's doctor of pharmacy. Today's pharmacist practices clinical pharmacy, delivering pharmaceutical care to patients in a variety of practice settings ranging from the familiar community pharmacy to lesser known but important settings such as hospitals, ambulatory clinics, and nursing homes. With all of the health professions evolving rapidly amid the backdrop of globalization, this book was written mindful of the important trend toward inter-professional education and cooperation among the various health care professions. An important step in understanding the uncertainties of the present and future is to understand the past from which it all came. As President Harry S. Truman once observed, "The only thing new in the world is the history you don't know."[13]

Special Features of the Book

Each chapter begins with an essential question for the reader to keep in mind as s/he reads the chapter. To supplement the main narrative, the book offers interesting sidebars in the form of box features and "pop-ups." These features combined with illustrations, charts, and diagrams provide you with a fuller picture of the history of pharmacy from multiple perspectives and the *zeitgeist* of a particular era. With students and educators in mind, at the end of each chapter there is a summary, a list of key words, and learning outcomes. The history of pharmacy has been an important part of my life and it is my sincere hope that you will enjoy it as much as I have.

Notes

1 George A. Bender, *Great Moments in Pharmacy* (Detroit, MI: Northwood Institute Press, 1966), 7.
2 David L. Cowen and William H. Helfand, *Pharmacy: An Illustrated History* (New York: Harry N. Abrams, Inc., 1990), 17.
3 Edward Parrish, *Proceedings of the American Pharmaceutical Association* (1866, 14: 77–8), 257–264.
4 G.P. Srivastava, *History of Indian Pharmacy, Vol. 1* (Calcutta: Pindars, Ltd., 1954), 125.
5 John Naisbitt and Patricia Aburdence, *Megatrends 2000* (New York: William Morrow & Company, 1990), 12.
6 William H. Zellmer, "Searching for the Soul of Pharmacy," *American Journal of Health-System Pharmacy*, Vol. 53 (1996), 1911–1916.
7 Hermann Peters, *The Pictorial History of Ancient Pharmacy with Sketches of Early Medical Practice*, trans. William Netter (Chicago, IL: G.P. Engelhard & Company, 1889), 5. (Reprint, Lexington, KT: Forgotten Books, 2011) www.forgottenbooks.com.
8 Rachel Naomi Remen, *My Grandfather's Blessings: Stories of Strength, Refuge, and Belonging* (New York: Riverhead Books, 2000), 28.
9 Ibid., 102.
10 Lewis Thomas, *The Youngest Science: Notes of a Medicine-Watcher* (New York: Viking Press, 1983), 200.
11 Norman Cousins, *Human Options* (New York: Norton, 1981), 214.
12 Lewis Thomas, *The Youngest Science*, 53.
13 Gorton Carruth and Eugene Ehrlich, eds., *American Quotations* (New York: Avenel Books, 1988), 275.

1 Prehistoric Pharmacy

From Stones and Bones to Weeds and Seeds

Belief begins where science leaves off and ends where science begins.[1]
(Rudolf Virchow)

The great astrophysicist Carl Sagan once wrote, "The world is very old and human beings are very young."[2] In a passage from *The Dragons of Eden* entitled "The Cosmic Calendar" Sagan made the complex comprehensible by taking all of history and compressing it into a single year. He placed the 15 billion-year history of the universe in context placing the origin of the universe on January 1, the formation of Earth on September 14, the origin of microbes on October 9, the appearance of dinosaurs on December 24, and the appearance of humans at 10:30 p.m. on December 31. So according to this cosmic calendar, each of us has an astounding 0.16 seconds of life on this planet!

Scientists estimate that the Earth began about 4.6 billion years ago, with the first forms of life appearing as microbes about 4 billion years ago. The first human ancestor *Ardipithecus ramidus* or "Ardi" appeared about 4.4 million years ago in Eastern Africa, according to the archaeological findings of Dr. Tim White and his team of researchers.[3] This means that microbes and some of the diseases they may cause appeared long before humans did. Disease is by far older than humankind.

The great challenge for researchers today who study the Prehistoric Era is the lack of a written record, so we must rely on other types of evidence that fossils and artifacts leave us. Anthropologists add to our understanding of prehistoric people by studying so called "primitive" societies that still exist in remote areas of the world. By observing "primitive" societies anthropologists have a living window into the past. Some religious and/or medical rituals that are still practiced have remained largely unchanged for millennia, such as the pilgrimages made by very ill people high into the lakes of the Andes Mountains in Peru, led by modern day shamans who appear timeless.

As early as 80,000 years ago, prehistoric people engraved drawings of plants and plant parts on bones and deer antlers, perhaps as a way of passing down their intimate knowledge of their environment to their descendants.

Table 1.1 Carl Sagan's Cosmic Calendar

January 1	Big Bang
May 1	Origin of the Milky Way Galaxy
September 9	Origin of the Solar System
September 14	Formation of the Earth
September 25	Origin of life on Earth
October 2	Formation of the oldest rocks on Earth
November 1	Invention of sex (by microorganisms)
November 12	Oldest fossil photosynthetic plants
November 15	*Eucaryptes* (first cells with nuclei flourish)
December 1	Oxygen forms on Earth
December 24	First dinosaurs
December 31, 10:30 p.m.	First humans
11:59:20 p.m.	Invention of agriculture
11:59:59 p.m.	Renaissance in Europe
January 1 (Second year) 12:00 a.m.	Widespread development of science, medicine, and technology

Additional fossil evidence of pharmaceutical knowledge has been found buried in the Shandar Cave in northern Iraq, where humans were buried with clusters of flowers and herbs, including marshmallow, yarrow, and groundsel.[4] Similarly, about 14,000 years ago, cave artists painted the image of a shaman or medicine man on the walls of the cave Les Trois Freres in Arriege, France.[5] He/she was depicted wearing the skin of a deer topped by a rack of antlers. Was he or she perhaps the first pharmacist?

One of the most exciting recent discoveries that have changed our thinking about our prehistoric ancestors was the discovery of "Ötzi," the 5,300-year-old "Iceman" whose remains were recovered frozen in a glacier in the Tyrolean Alps. This find was remarkable for a number of reasons. The Iceman was hunting in the Alps at a high altitude when he was wounded and overcome by a massive snowstorm, which preserved his remains for several millennia. Ötzi carried a utility belt wrapped around his waist that contained pouches of dried mushrooms and fungi that had antibacterial properties. Ötzi also had a number of tattoos on his body that corresponded to acupuncture points, which led researchers to believe that Neolithic people had a much more sophisticated knowledge of the body and of herbs than previously thought.[6] Apparently, some of our Neolithic ancestors understood acupuncture. This is not surprising, since our *Homo sapiens* ancestors share the same brain capacity and innate curiosity that we do, but unfortunately did not have the benefit of millions of years of human discovery from which to benefit as we do. One of the important lessons of history is empathy, that is, to place one's self in the historical circumstances of our ancestors to understand their motives, actions, and accomplishments. It is important to bear in mind that we are here today due in large part to their individual and collective

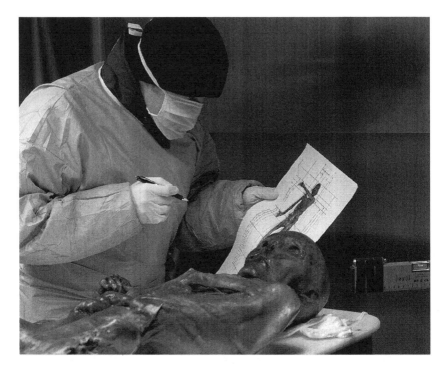

Figure 1.1 Picture of Ötzi—Ötzi (the 5300 year old Iceman) was discovered in the Ötztal Alps on September 19, 1991. Ötzi's discovery yielded a wealth of information about Neolithic medicine.

Source: © South Tyrol Museum of Archaeology—www.iceman.it.

ingenuity, while most other species that have inhabited this planet have long since passed into extinction.

The Paleolithic Era

The Prehistoric Era is customarily divided into two periods: the Paleolithic or Old Stone Age, and the Neolithic or New Stone Age. Life in Paleolithic societies, where people lived in caves and survived by hunting and gathering food, dates back to about 40,000 BCE. Paleolithic people lived in low-density populations in small clans or tribes of less than 100 people and moved from place to place in search of food. Human life at this time was precarious, uncertain, and relatively brief in duration. Any disease or injury might well have meant death for those afflicted. Until 1800 CE the overall life expectancy was about 25 years with some notable exceptions such as during Roman times. This average has been adversely affected by extraordinarily high infant mortality rates for this time period.

Despite leading lives governed by necessity, Paleolithic people used their instincts as well as simple trial and error to learn how to cope with their environment. For example, when someone is stung by a bee, a natural reaction is to rub the afflicted area with something cool and grainy such as mud, which brings relief from the pain and also has the potential to help remove the source of the pain, the bee's stinger. Further, people observed what animals ingested and also learned by trial and error what was helpful in healing and what was not. Over time, individuals in a tribe emerged who had special talents for healing others, known as "shamans," "witch doctors," or "medicine men." The term "shaman" derives from the Siberian language meaning "one who knows."[7] Thus, the role of the learned and trusted healer in the person of the shaman evolved over time into the modern physician, pharmacist, nurse, and other health care professionals.

Paleolithic people explained the world around them in supernatural or religious terms. In fact, from Paleolithic times through the Renaissance Era, cultures and civilizations would explain illness in sacred or religious language. It would not be until well into the Early Modern Era when pharmacy and medicine would be framed in secular or scientific language. Shamans developed theories of illness based on supernatural causation. Causes of illness might include a person who broke a tribal taboo, dishonored a god, or had a spell cast against them by another shaman. In order to heal the afflicted person, the shaman enlisted the whole tribe in the healing process, and the patient became a symbol of this conflict between illness and health. The patient would "confess" (today we use the term "present") their symptoms to the shaman and the shaman would assess the problem and offer a cure. Even with all of our modern technology, 75 percent of all primary diagnoses derive from the patient's presentation of symptoms.[8] It is important to note that the shaman treated the "whole" patient. The mind, body, and spirit were treated as one entity and not compartmentalized as they often are in modern health care.

The goal of the shaman was to cast out the "evil" or purify the patient by means of both oral and manual rites. Oral rites were symbolic and palliative in nature and included chants, incantations, and dances. By contrast, manual rites were physical actions performed on the patient by the shaman and included, performing a rudimentary surgical procedure, sucking out poison from a wound, or administering herbs for a patient to smell or swallow. Indeed, administering herbs or potions might well have been the first form of drug therapy. Depending on the severity of the illness or local custom the shaman might administer an especially harsh potion. *Dreckapotheke* is a term derived from medieval German slang used to describe an especially disgusting concoction or fumigation used by shamans to ward off evil.[9] Even in Paleolithic times, people respected the power of medications to fight disease and saw the need for them to be strictly controlled and administered by a relatively learned and trusted individual. The shaman's cauldron became a symbol of magic, mystery, and transformative power that still lingers in modern memory in the symbol most closely associated with pharmacy—the mortar and pestle.

Mortar and Pestle

The mortar and pestle are perhaps the oldest known tools associated with early pharmaceutical work that have come to be recognized together as the icon for pharmacy itself. The mortar's name derives from the Latin *mortarium*, meaning "a receptacle made of a hard material" as well as "the product of grinding." The pestle's name derives from the Latin word *pistillum*, meaning "a club shaped hand tool for grinding substances."[10] As archaeological artifacts they date back at least 6,000 years and have been found all over the world. Mortars and pestles were tools used to grind crude herbs into powder. The earliest specimens were made of stone or wood and became more elaborate as humans became more skilled at metallurgy. The Wedgewood mortar and pestle from 1779 became popular because the mortar was made of porcelain, and the pestle had a wooden handle and porcelain bottom. Similarly, the ubiquitous mortar and pestle that are found in nearly every pharmacy today are both made of porcelain.

The modern reader rightfully might be asking, did this treatment work? Modern medicine tells us that most physical human illness is self-limiting, so most of the time the patients recovered regardless of what the shaman did. Still, modern medicine is becoming more intrigued with the power of the "placebo" effect. The word "placebo" derives from Latin *placere* and literally means "I shall please."[11] Studies have repeatedly shown that if a patient has confidence in the treatment and the person administering it, the chances for recovery increase.[12] In Paleolithic times shamans performed brain surgery known as trepanation to relieve pressure on the skull caused by head trauma. Archaeologists have uncovered human remains that indicate that 75–80 percent of the patients on which a trepanation was performed survived this procedure. At first it was believed that trepanation was a religious ritual, or perhaps a sacrificial act, but the fact that bone had re-grown smoothly around the incisions proved that the patients survived the trepanation for some time after the procedures were performed. Some skulls demonstrated that some patients survived multiple trepanations. Shamans performed trepanations to release the evil spirits they believed were inside their patients' heads causing pain. Using sharpened stone tools, the shaman made the incision on the top of the patient's skull in order to let the demons escape. A piece of bone was cut out and would be given to the patient after the trepanation to wear as an amulet to ward off future demons from attacking. Archaeological evidence

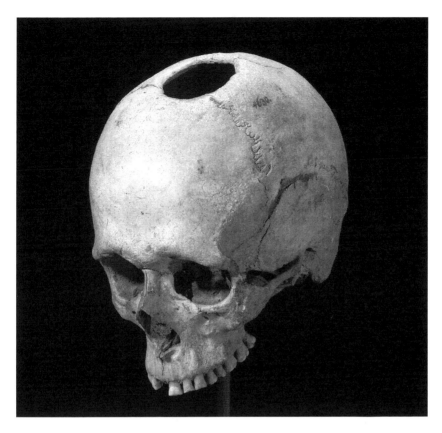

Figure 1.2 Trepanned Skull—This prehistoric trepanned skull revealed that our ancestors performed brain surgery using stone instruments and that the patients survived more than 75 percent of the time.

Source: Nationalmuseet/National Museum of Denmark.

shows that trepanations occurred all around the world. Even today in some parts of the world, trepanations are still performed much as they had been in prehistoric times, and for similar reasons. In addition to trepanations, archaeological evidence indicates that shamans also knew how to set bone fractures, perform cauterizations, and perform amputation of limbs.[13]

Neolithic Age

About 8,000 years ago, something remarkable occurred in human history: people learned how to farm and domesticate animals, which allowed them to settle in higher density populations in more permanent settlements. With higher density population came political hierarchy, division of labor, and

Materia medica means anything from the natural world that can be used to heal a patient: plant, animal or mineral. The term *materia medica* was coined by the great ancient Greek physician and botanist Pedanios Dioscorides.

specialization that contributed to the growth of shamans who had the time and means to expand their knowledge of the *materia medica* around them.

To be sure, the Neolithic Age and the Agricultural Revolution that accompanied it represented a major advancement for humanity, but it also brought about some unintended consequences. For example, the domestication of animals meant a convenient and stable food supply, but with many people living in close proximity to many animals, it also created an environment for cross-species transmission of infectious diseases. None other than Dr. Edward Jenner observed:

> The deviation of man from the state in which he was originally placed by nature seems to have proved to him a prolific source of diseases. From the love of splendor, from the indulgences in luxury, and from his fondness for amusement he has familiarized himself with a great number of animals, which may not have been intended for his associates.[14]

Humanity's association with cattle brought us smallpox and mad cow disease. In return, humans brought tuberculosis to cattle. Pigs and fowl threatened us with influenzas of varying types and degrees. Dogs brought us measles. Dogs, cats, chickens, mice, rats, and reptiles brought us salmonella. Higher density human and animal populations also meant higher levels of animal fecal waste in water supplies, which then harbored diseases including polio, typhoid, viral hepatitis, whooping cough, and diphtheria, to name a few. Indeed, history teaches us about continuity and change, and that neither comes without a price in terms of what has been lost, sometimes forever. The price of the Agricultural Revolution and the Neolithic Age came in the form of increased risk of contracting infectious disease and the advent of warfare because of competition for fixed territorial resources.[15] Despite our advanced technology we are still at the mercy of the common cold, and suffer through them much the same as our prehistoric ancestors did.

The prehistoric drug legacy that reaches across thousands of millennia is a powerful one. Today, about 25 percent of all prescription drugs are derived from trees, shrubs, or herbs. The figure rises closer to 50 percent when drugs made from microorganisms are included.[16] According to the World Health Organization, of 119 plant-derived pharmaceutical medicines, almost 74 percent are used in ways that correlate directly with their traditional uses as plant medicines by indigenous cultures. Astonishingly, less than 10 percent of the world's plants have been analyzed and studied for their possible therapeutic properties.[17]

Chapter Summary

This chapter places the history of the universe in a historical framework and explains how our prehistoric ancestors explained illness and injury and how they treated themselves. The chapter also shares the latest research into how prehistoric people knew so much about their environment and experimented with plants and fungi that had antibacterial properties. The rise of the shaman as healer is explored.

Key Terms

Ardi	oral rites	trepanation
Ötzi	manual rites	amulet
Paleolithic	*dreckapotheke*	*materia medica*
Neolithic	mortar and pestle	Agricultural Revolution
shaman	placebo	

Chapter in Review

1 Do you understand the prevailing theory about how the universe started and how relatively late human beings appeared on the Earth?
2 Explain in detail how early humans viewed disease and treated it during the Paleolithic and Neolithic Eras.
3 Explain the importance of the shaman's rise as a healer in prehistory and his connection to modern health care.

Notes

1 Fielding H. Garrison, *An Introduction to the History of Medicine* (Philadelphia: W.B. Saunders Company, 1929), 14.
2 Carl Sagan, *The Dragons of Eden* (New York: Random House, 1977), 13–17. For a more detailed graphic explanation see "The Cosmic Time-Line," http://visav.phys.uvic.ca/~babul/AstroCourses/P303/BB-slide.htm.
3 Yohannes Haile-Selassie and Giday Wolde Gabriel, eds., *Ardipithecus Kadabba: Late Miocene Evidence From the Middle Awash Ethiopia* (Berkeley: University of California Press, 2009), 159.
4 David L. Cowen and William H. Helfand, *Pharmacy: An Illustrated History* (New York: Harry N. Abrams, 1990), 17.
5 Barbara Griggs, *Green Pharmacy: The History and Evolution of Western Herbal Medicine* (Rochester, VT: Healing Arts Press, 1997), 1. Albert S. Lyons and R. Joseph Petrucelli, II, *Medicine: An Illustrated History* (New York: Abradale Press, 1987), 27. S.E. Masengill, *A Sketch of Medicine and Pharmacy* (Bristol, TN: The S.E. Masengill Company, n.d.), 7.

Roy Porter, *The Greatest Benefit to Mankind: A Medical History of Humanity* (New York: Norton Books, 1997), 31. Victor Robinson, *The Story of Medicine*. 1931. Reprint. New York: The New Home Library, 1944), 6.

6 Konrad Spindler, *The Man in the Ice: The Preserved Body of a Neolithic Man Reveals the Secrets of the Stone Age*, trans. Ewald Osers (London: Phoenix Books, 2001), passim. Michael T. Kennedy, *A Brief History of Disease, Science, and Medicine* (Mission Viejo, CA: Askelpiad Press, 2004), 1–3. Sally Pointer, *The Artifice of Beauty* (Gloucestershire: Sutton publishing, Ltd., 2005), 7.

7 *The American Heritage Dictionary*, 1973 ed., s.v. "shaman."

8 "The New Medicine," dir. and prod. Gerald Richman, PBS, Twin Cities Public Television/Middlemarch Films, Inc., 9 April 2006.

9 Roy Porter, *The Greatest Benefit to Mankind: A Medical History of Humanity* (New York: Norton Books, 1997), 35.

10 *The American Heritage Dictionary*, 1973 ed., s.v. 855, 980.

11 *The American Heritage Dictionary*, 1973 ed., s.v. "placebo."

12 "Ancient Evidence: The Real Disciples and the Placebo Effect," dir. and prod. Gillian Bancroft, Discovery Channel, 14 January 1998. Lewis Thomas, *The Youngest Science* (New York: Viking Press, 1983), 15. *The St. Louis Post-Dispatch*, "Placebos are Regularly Used by 50% of Doctors," 24 October 2008, sec. A: 20. *The St. Louis Post-Dispatch*, "The 'Placebo Effect' May Not Require Deception," 23 December 2010, sec. A: 21.

13 Lois Magner, *A History of Medicine* (New York: Marcel Dekker, Inc., 1992), 6. Fielding H. Garrison, *An Introduction to the History of Medicine*, 28. Roy Porter, *The Greatest Benefit to Mankind*, 35. Ann Rooney, *The Story of Medicine* (London: Arcturus Publishing, 2009), 130–131.

14 Roy Porter, *The Greatest Benefit to Mankind*, 19.

15 Jared Diamond, "The Worst Mistake in the History of the Human Race," *Discover Magazine*, May 1987: 64–66.

16 N.R. Farnsworth, O. Akerlee, A.S. Bingel, D.D. Soejarto, "Medicinal Plants in Therapy," *Bulletin of the World Health Organization*, 63, no. 6 (1985): 965–981.

17 Diarmuid Jeffreys, *Aspirin: The Remarkable Story of a Wonder Drug* (New York: Bloomsbury Publishing, 2004), 10.

2 Ancient Pharmacy in the River Civilizations of Mesopotamia and Egypt

> We are so often tempted to believe that our way of doing things is not only better but also different than those of times long past, that we do not only always realize how closely our methods resemble those practiced during civilizations which are dead and almost forgotten.[1]
>
> (Otto A. Wall, 1898)

The Ancient River Civilizations

How did early civilized people view health and illness?

The Neolithic Era (8000–3000 BCE) became a victim of its own success. Over time Neolithic villages grew in population and those near rivers in fertile regions of the world thrived. With increased prosperity and the ability to grow and store food in granaries, Neolithic villages expanded, annexed, or conquered others. As early as 4000 BCE in some regions of the world, cities and civilizations emerged with the ability to produce written records. This ushered in the era of the Ancient World and recorded history. Ancient river-based civilizations emerged in Mesopotamia, Egypt, India, and China. Although each of these civilizations would develop its own mythologies and world views, they all relied on similar highly centralized hierarchies with strong priestly castes who exercised tight control over the world of ideas and practices in each of their respective civilizations. To be sure, one goal of these early ancient civilizations was to produce strict conformity and monolithic thinking among their people. In terms of medicine and pharmacy, these priestly castes were literate and learned men who shaped the theories of illness and practices for treating patients within their respective civilizations. The beliefs and practices of these early ancient civilizations might appear quite harsh to modern sensibilities, but no one understood better than the people living in these ancient civilizations that they were one step away from the prehistoric barbarism that surrounded them. Closer to our own time, Sigmund Freud warned us that, "civilized society is perpetually threatened with disintegration."[2]

Ancient Mesopotamia

One of the first civilizations to emerge from the shadows of the Neolithic Era was Mesopotamia, which was located between the Tigris and Euphrates rivers in what is mostly modern day Iraq. This area, known as the Fertile Crescent, became an ancient crossroad of marauding civilizations that competed for its fertile land and resources. Although the basic foundations and beliefs of Mesopotamian civilization were established by an inventive people known as the Sumerians (4000–2000 BCE), they would fall victim to the Babylonians (2000–1350 BCE), the Assyrians (1350–612 BCE), and the Chaldeans (612–539 BCE). The Sumerians wrote in a pictogram form of writing known as cuneiform on wet clay tablets that were baked in kilns to make them permanent documents. In addition to cuneiform, the Sumerians developed basic arithmetic, as well as a standard system of weights and measures based on a sexagesimal system. Our system of recording seconds and minutes based on 60 is a vestige of this ancient system. As many early civilizations did, the Sumerians placed great emphasis on astrology as a macrocosm of the universe and viewed the human body and how it worked as a microcosm of that universe. Seasonal changes and how they affected the body were especially intriguing to the Sumerians. Astrology and the search for positive and negative omens played a key role in a priest-physician's decision of when to treat a patient. For example, a priest would avoid treating patients on the days of an illness that were divisible by seven due to the demons known as the Evil Seven.[3]

What Does Ancient Mesopotamian Astrology Have to Do with the Modern Flu?

The ancient Mesopotamians were astute astrologers, leaving us with over 7,000 clay tablets recording their observations. For them, human fate was determined by the position of the stars and planets. The Romans were also fascinated by astrology. Even today our words "unfortunate," "ill-starred," and "unlucky" remain as vestiges of this ancient belief. The Italian word *influenza* was derived from the Latin and means influence. Our modern word "influenza" derives from this ancient legacy and refers back to the streaming of ethereal fluid from the stars into people, an ancient belief explaining why some people contract the flu while others do not. That's fate for you![4]

Mesopotamian Mythology and the Caduceus

The Mesopotamians developed an intricate mythology to explain their world. They developed hierarchies of gods and goddesses who interacted with humans in both positive and negative ways capable of inflicting and curing disease.

Ea was the Babylonian god of water and the first god of physicians. Ea's son Marduk begot Nabu, who reigned over medicine. He was assisted by Gula, the goddess of death and healing, who became known as the Great Lady of Physicians.[5] In fact, there were several temples dedicated to Gula where patients apparently went to be diagnosed. The messenger god Ningishrida, who was a healing god, was depicted carrying a staff entwined with a double-headed snake, the ancient symbol of healing that predates our modern symbol for medicine—the caduceus. In the Sumerian creation story, *The Epic of Gilgamesh*, a snake stole the plant of eternal life and thus the serpent became a symbol of regeneration and healing in the ancient world.[6] By contrast, the Mesopotamians also had a pantheon of demonic gods and goddesses who inflicted illness upon humans. Nergal was the god of death and destruction who was aided by the god of plague Nasutar and Pazuzu the goddess of sickness.[7] They, in turn, were assisted by lesser known demons who were associated with specific diseases or symptoms and included: Ashakku (consumption), Tiu (headache), Namtaru (sore throat), Axaxuzu (jaundice), and others.

The Mesopotamians viewed their world in supernatural terms and it was not surprising that their theory of illness closely resembled those of prehistoric times. Disease could be inflicted upon a person by fate, personal carelessness, irreverence, impiety, spiritual impurity, or personal weakness. Although all of the Mesopotamian healers were first and foremost trained as priests, they specialized in various aspects of healing patients in mind, body, and spirit—they treated the whole patient. The *baru* was the seer/diviner, operating in the spiritual realm, whose job was to diagnose a patient and offer a prognosis. They searched for omens, often by sacrificing a lamb and examining its liver. This augury took the form of hepatoscopy, or the removal of the lamb's liver. For the Mesopotamians, the liver was the seat of life and the central organ in the body where the blood collected.

Mesopotamian Practitioners

The task of the *ashipu* was to drive out the evil demons from the patient's body, home, and village through oral rites. Some scholars have uncovered records of *ashipus* who prescribed drugs for patients. The *asu* was the physician/priest who functioned on a spiritual level, but also

Weighing about five pounds, the liver is the largest gland in the body and secretes bile that accumulates in the gall bladder. Centrally located in the body, the Mesopotamians believed it was the central organ of the body where the soul resided. Pig's gall or bile was used for ailments of the eye and to treat anal growths in ancient Egypt. Later, the Greek physician/botanist Dioscorides also recommended pig's gall to treat eye ailments. In fact, as late as the 1930s ox gall, known as the extract *Fellis Bovini*, was still officially listed in the British Pharmacopoeia.[8]

Figure 2.1 Augury Tablet—Augury tablets were stone depictions of animal livers that were used to train healers on how to read the liver for omens to offer a prognosis to their patients.

offered drugs and performed surgery. While the *baru*, *ashipu*, and *asu* primarily treated the royal court and upper classes, there were barbers who performed some rudimentary dentistry and surgery on the lower classes.[9] Interestingly, the Mesopotamians believed that tooth decay was caused by worms gnawing away at teeth. This belief persisted until the eighteenth century just about the time Pierre Fauchard (1678–1761) pioneered modern dentistry.[10]

According to the 32,000 tablet library that has survived at Nineveh, 800 of those tablets dealt with Mesopotamian *materia medica*. One of the tablets dates back to 2100 BCE and contains 15 drug prescriptions in what might well be the first herbal *materia medica*.[11] Curiously, although the Mesopotamians had a standard system of weights and measures, no indications remain that they used them when compounding drugs.[12] These tablets list 250 drugs of plant origin and 120 of mineral origin, as well as 30 simple unmixed drugs listed for making compounds, e.g. mandrake, opium, hellebore, pine turpentine, licorice root, myrrh, asafoetida, mentha, mustard, cannabis, crocus, turmeric, thymus, honey, cedar, willow (a precursor to modern aspirin), and others.[13] They also used the chemical substances alum, arsenic, and sulfur.[14] Later, the Babylonians added senna, saffron, coriander, cinnamon, and garlic to the Sumerian collection of herbs. These herbs were often mixed together into what are known as poly-pharmaceuticals, a practice that would endure until the nineteenth century.[15]

Oils, alcohols, wines, fats, milk, wax, and beer were used as vehicles to administer these drugs. Some records indicate that the Mesopotamians would devote as much as one-third of their annual grain harvest for brewing beer. They knew about natural fermentation as early as 2500 BCE.

In fact, the Mesopotamians discovered the process of distillation which allowed them to make essence of cedar and other oils.[16] Interesting sets of compounds were produced by mixing olive oil and beer as a shampoo, and using warm oil to soothe earache.[17] For example, one prescription for a poultice read:

Sift and knead together—all in one-turtle shell, the sprouting naga plant, salt and mustard; wash with quality beer and hot water; scrub (the sick spot) with all of it; after scrubbing, rub with vegetable oil and cover with pulverized fir.[18]

At first glance, the modern reader might view this remedy as ridiculous, but upon closer examination it makes medical sense. The pulverized turtle shell and salt have anti-bacterial properties, and washing the wound with beer that contains alcohol and will rid it of foreign matter starts to make sense to us. After cleaning the wound, they applied vegetable oil and pulverized fir shavings to cover the wound to promote healing much in the way we might use a bandage.

Mesopotamian Dosage Forms

The archaeological record also indicates that a "pharmacy street" existed in the city of Sippur where people sold medicinal drugs.[19] Salt and beer were used as antiseptics and saltpeter was used as an astringent to harden skin. Mesopotamian priest-physicians washed wounds and applied poultices and bandages.[20] Some of the earliest poultices were composed of rotten grain and water, while others were made of a powdered substance called *siklu* which was a paste made from garlic or onion kneaded with cassia juice. Another poultice was made from mashed turnips mixed in milk. Mustard plasters or poultices made from mustard mixed with flour and water were also used. These poultices were applied to the skin and held in place with a bandage.[21] In addition to poultices, the Mesopotamians offered various dosage forms including: decoctions (boiling down to extract flavor or to concentrate), electuaries (herbs mixed in sugar water or honey), embrocations (liniments), enemas, fumigations, lotions, medicated wines, ointments, and plasters.[22]

Figure 2.2 The World's Oldest *Rx*—The world's oldest prescription tablet (*c.*2200 BCE) from Mesopotamia indicates that medical treatment was not entirely based on magic.

Source: Courtesy of Penn Museum, image 150014.

Mesopotamian Cosmetics

In addition to drugs, the Mesopotamians introduced cosmetics to the world. Both men and women used a variety of cosmetic pigments in a wide array of colors including blue, purple, red, yellow, green, and black. Archaeologists have discovered these pigments stored in seashells. The green and blue pigments were made from copper compounds; the reds from iron oxide; and the black from manganese oxide. These raw colors were lightened by mixing them with a white pigment composed of lead carbonate (ceruse) and apatite (calcium phosphate). Apatite was derived from burning bones and was used as a cos-

> The Mesopotamians used hair oils to style their hair and one recipe for a hair dye was made from leeks and cassia extracts.[23]

metic pigment from prehistoric times until the late nineteenth century.[24]

Hammurabi's Code, the oldest written law code, indicates that surgery was regulated. The treatment of wounds, broken bones, tumors, and abscesses was also noted in the laws, and if the operations failed, priest-physicians faced harsh penalties such as having their hands cut off, or in rare cases death. Surely, such harsh penalties might have discouraged practitioners, but nonetheless the number of practitioners continued to grow over time as did the number of surgeries, suggesting that perhaps the penalties were rarely enforced to the letter of the law. The archaeological record has also yielded a number of surgical instruments that were used. By virtue of Mesopotamia's geographic position as a crossroads among the continents and its pioneering efforts in medicine and pharmacy, its influence would be felt for millennia in the human quest for health and well-being.

Ancient Egypt

How did the Ancient Egyptians advance pharmacy and medicine?

At about the time Mesopotamia was forming, another great river civilization emerged from its prehistoric shadow—Egypt. "Egypt" literally means "the gift of the Nile." Similar to Mesopotamia, Egypt was ruled by a central monarch known as a pharaoh, who was assisted by a powerful ruling class of priests and bureaucrats. The pharaohs ruled Egypt as gods, and wielded absolute power. The Egyptians also had an elaborate polytheistic mythology that guided every aspect of their lives, including pharmacy and medicine. The Egyptians valued specialization in all vocations, especially in medicine and pharmacy. This was reflected in their pantheon of gods in which pharmacy was separated from medicine. Thoth was the creator of science and medicine, who reigned as the patron god of priest-physicians and battled against Sekmet, the goddess of pestilence. Anubis was the trusted apothecary to the gods who prepared prescriptions and was in charge of the storehouse of medicines. He also served as the keeper of the chamber where mummies were embalmed. The goddess Isis, who symbolized the earth mother and possessed great knowledge about herbs and poisons, served as the herbalist to the gods. It is important to note here that from the start the Egyptians placed great significance on providing a designated place for pharmacy in their medical system, in which drugs would play a key role.[25]

Imhotep

Egypt gave us the first great physician in history that was known by name. Imhotep (*c*.2625 BCE), which literally means "One who comes in peace," served as grand vizier to the Pharaoh Zoser and was a man of remarkable talent who designed the step pyramid at Saqqara. He served as the high priest to the Sun god Ra. Physician, architect, vizier, artist, sculptor, and priest, he was by all accounts a universal genius who was one of the only mortals to become deified as a god of medicine due to his amazing contributions to humanity. Some speculate that he might have written what became the Edwin Smith papyrus. Imhotep became the patron god of scribes, who drank a toast to him prior to beginning a writing session. By the sixth century BCE, Imhotep replaced Thoth as the god of medicine and became the model for his ancient Greek counterpart Asclepius. Imhotep remains a legendary figure because his tomb has yet to be discovered.[26]

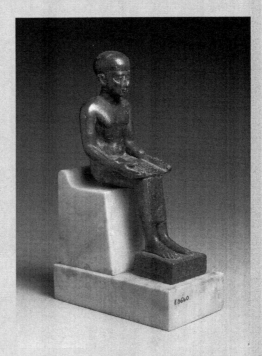

Figure 2.3 Statue of Imhotep (*c*.2600 BCE). Imhotep was a polymath who served as a vizier and a healer. He was one of the first mortals to later be deified for his medical expertize in ancient Egypt.

Source: Photo: Hervé Lewandowski. Musée du Louvre. © RMN–Grand Palais/Art Resource, NY.

Ancient Egyptian Theory of Illness

Similar to the Mesopotamians, the Egyptian theory of illness was explained in supernatural terms. The sick person had become out of balance with the cosmos by their own actions, such as disrespecting the gods, the dead, or spirits. Illness also might have been caused by the actions of demons. Therefore the treatment was to restore the body's balance with the cosmos by oral and manual rites. As close observers of their beloved river, the Egyptians believed that just as the Nile had tributaries branching from it, so too did the body, and they called human blood vessels *metu*. For the Egyptians, the seat of life was the heart, the place where all of the bodily canals or *metu* met and the site of respiration. In fact, the heart was so highly regarded by the Egyptians that during mummification rituals it was often one of the few internal organs to be carefully removed and preserved. By contrast, the brain was simply removed from the skull by scraping it out in pieces with a hook through the nasal cavity. The Egyptians believed that four elements flowed throughout the body, including blood, air, water, and *wekhudu* (bodily waste). For the Egyptians, excessive *wekhudu* was the cause of illness and thus the primary focus of their preventive medicine and treatment was to purge the body of this waste.[27]

"Mummy Powder"

One of the most bizarre ingredients ever used in drug preparation was the ground-up remains of human mummies in the form of mummy powder. The use of mummy powder as a treatment for conditions ranging from paralysis to contusions dates back to at least the eleventh century and flourished all the way to the end of the eighteenth century. Reliable sources such as Avicenna, Ambrose Pare, and Pierre Pomet mention its use in their writings over these centuries. Pomet, who served as King Louis XIV of France's personal druggist, classified five types of mummies that were used to make the coveted powder, and ranked them in order of preference: the true Egyptian, the factitious that were embalmed with bitumen and pitch, the Arabian embalmed with myrrh and aloes, bodies dried in the sun along the Libyan coast, and artificial mummies that were made of people who had been recently killed and mummified. Apparently, as the supply of genuine ancient Egyptian mummies dwindled, they were supplemented by other forms of mummies made to order.[28]

The Birth of Modern Egyptology

For almost a millennium, our knowledge about ancient Egypt came to us from what the Greeks and Romans wrote about them. From the fifth century CE until the late eighteenth century the Egyptians' form of writing, called hieroglyphics, could not be deciphered until a talented French linguist Jean François Champollion (1790–1832), literally broke the code embedded in the Rosetta Stone. The Rosetta Stone had been discovered in 1799 by one of Napoleon's soldiers, and was sent to France for study. Champollion completed the yeoman's task of taking the ancient Greek inscriptions and comparing them to the hieroglyphic pictograms, thus unlocking the world of ancient Egypt for us. This discovery ushered in the field of Egyptology. The Egyptians left a rich legacy of information about their civilization written in hieroglyphics on papyrus scrolls and in stone. Among the most detailed of the papyri were 11 concerning pharmacy and medicine. The Egyptians were especially detailed in their descriptions of the drugs they used. These were outlined in the Ebers papyrus, named in honor of the German Egyptologist Georg Ebers. The Ebers papyrus was written about 1550 BCE, when Egyptian medicine was at its closest semblance to rational medicine, although as each compound was prepared, prayers and incantations were believed to imbue the formulas with their healing power. For the ancient Egyptians it was still the magic, rather than the actual drugs that possessed the power of healing.[29]

The Ebers Papyrus

The Ebers papyrus was largely a formulary that outlined 811 drug recipes composed from 700 drug ingredients derived from plants, animals, and minerals. The drug ingredients derived from plants included acacia, almond, aloe, aniseed, castor bean, celery, cinnamon, date, dill, fennel, fig, garlic, hemp dogbane, henbane, juniper, lettuce, lotus, mandragora, myrrh, onion, rotten bread (primitive penicillin?), peppermint, poppy seed, saffron (the blood of Thoth), senna, tamarisk, vervain (the tears of Isis), watermelon, willow (*tjeret*), and wormwood. To treat diabetes or "the wasting disease," the Egyptians applied mud and moldy bread over sores to keep them from becoming infected. The drug ingredients derived from animals included pig's brain, fly's dirt (excrement), crocodile dung (used in powder form as a contraceptive), and hippopotamus fat. Mineral ingredients included alum, copper, iron oxide, limestone, sodium carbonate, salt, and sulfur.[30]

    The Ebers papyrus also demonstrated that the Egyptians also had an impressive knowledge of dosage forms including gargles, snuffs, inhalations, suppositories, fumigations, enemas, poultices, decoctions, infusions, troches (lozenges), lotions, ointments, and plasters.[31] The Egyptians also used catheters.[32] Unlike the Mesopotamians, the Egyptians used quantitative formulas in preparing their poly-pharmaceutical compounds. The Egyptians have left us evidence that they used mortars made of stone, hand mills, sieves, and balances in preparing formulas. They stored their drugs in containers made of pottery and glass.[33] The archaeological record also indicates that the Egyptians were highly specialized in their division of labor when it came to the preparation of drugs. In the royal court there was a special room in the temple known as the *asi-t* that was designated for the preparation of drugs under the supervision of a Conservator of Drugs, and a Chief Preparer of Drugs who also served as the Chief of Royal Physicians. Assisting these officials were a secondary tier of specialists known as drug collectors, drug preparers (*pasto-phors*), and drug conservers. The place where drugs were stored was called the Royal Warehouse and the royal laboratory was known as the House of Life.[34] So, in large measure the ancient Egyptian priest-physician relied very heavily on drugs to treat patients and performed the tasks of modern pharmacists. More remarkably, all of these priest-physicians and all health care operations in ancient Egypt were publicly financed by the pharaoh's theocracy.

Ancient Egyptian Cosmetics

The Egyptians, like the Mesopotamians, also developed an impressive line of cosmetics which were used by both men and women. Geography, and later vanity, played a role in the development of Egyptian cosmetics. Sunscreen became a necessity for people working under the scorching African sun, so much so that in the entire history of the construction of the pyramids the only incident involving a labor dispute was when the authorities ran out of sunscreen. The strike was soon settled after sunscreen was delivered to the workmen and they returned to work. The most common forms of sunscreens were made from sesame and castor oils. More elaborate sunscreens were made from various aromatic plants such as the blue Nile lotus that was mixed with oil. Sunscreen soon led to formulas for make-up foundation, rouges, lipsticks, and eye shadows composed of copper, antimony, and charcoal. The classic Egyptian look, featuring dark eye shadow, played a practical role of acting as a sun-block, protecting the area around the eyes from sunburn.[35]

Due to the intense heat and religious rituals calling for routine bathing and personal hygiene, the Egyptians practiced sound public health practices. According to the Ebers papyrus, sand or pastes made from ashes, clay, chalk, alabaster, natron (salt), and honey were used as exfoliant scrubs to cleanse the skin before bathing. They used facemasks made from beaten egg whites and scented oils to moisturize the skin. Prior to each daily meal, they washed themselves in alkali solutions.[36]

Many Egyptians also believed that hairlessness was next to godliness. They shaved using bronze blades to remove their body hair. They satisfied their need for hair with wigs. The finest wigs were made from human hair while other wigs were made from date palm fiber or from hay. They also fashioned false beards which accounts for why the female pharaoh Hatshepsut was depicted sporting a beard. Perfumes were developed, consisting of aromatic gels, pressed into cones, which were worn on top of their wigs, releasing perfume as they melted.[37]

Making Ancient Egyptian Perfumes

The ancient Egyptians were masters at making perfumes. The simplest and perhaps earliest method was to place aromatic plants in jars that were heated to capture the aromatic steam with a cloth placed over the jar. The cloth was then wrung out to capture the aromatic oil. Over time, perfume-making became more complex. Aromatic wood shavings were soaked in wine to elicit the scent and then the scented wine would be mixed with animal fat and simmered slowly over low heat. Once the fat gained the scent from the herbal wine, it was skimmed off and the product would be allowed to cool. Additional powdered herbs and spices were added to the product that would add the final, desired scents to it. The product would then be simmered slowly once again and cleansed of the plant residue and the finished perfume oil would be skimmed off the top for use. Frankincense and myrrh were popular aromatic resins.[38] For example, Cleopatra's favorite perfume was cyprinum composed of camphire from the henna plant.[39] Later civilizations would advance perfumes and cosmetics but later innovations were simply variations on the basic principles that the ancient Egyptians had established thousands of years ago.

The Birth of Laxatives

Egyptians from all social classes engaged in purification and personal hygiene rituals that have their origin in the much admired ibis, a bird native to the Nile River. In fact, the ibis was so revered that Thoth, the god of medicine, was depicted with the head of the ibis bird on a human body. The Egyptians admired the ibis' penchant for cleanliness through its meticulous preening of itself including the squirting of water into its anus to cleanse itself. Thus, the laxative, the suppository, and enema were born. Consistent with their belief in the *metu* as a system of internal bodily canals the Egyptians believed that waste materials collected near the anus, which if allowed to build up would cause disease if not eliminated from the body. On a monthly basis ancient Egyptians purged themselves by vomiting and taking laxatives including castor oil, colocynth, and senna. Suppositories and enemas were also used.[40] In fact, one priest-physician, Iri, was known as the Keeper of the Royal Rectum or the pharaoh's enema expert.[41]

The Smith Papyrus

Medical knowledge from ancient Egypt derives mostly from the Smith papyrus named in honor of the American Egyptologist Edwin Smith. The Smith papyrus dates back to about 1650 BCE and shows that similar to pharmacy Egyptian priest-physicians also specialized in treating various parts of the body. While there is no indication that any thoracic surgery was performed, Egyptian priest-surgeons set bone fractures, performed circumcisions, lanced boils, removed cysts, and excised tumors successfully. The archaeological record has yielded the discovery of ancient scalpels, knives, forceps, and probes. The Egyptians also cauterized wounds using red hot irons.[42] Although the emphasis in healing patients was placed on the supernatural, Egyptian priest-physicians were quite pragmatic in classifying illness into three categories that would guide their treatment plans. The optimal diagnosis for a patient was, "An ailment I shall treat." For chronic conditions the priest-physician offered a diagnosis that read, "An ailment with which I shall contend." For terminal cases the priest-physician sometimes recognized that a patient had, "An ailment not to be treated." In the terminal cases, priest-physicians offered palliative care for their patients.

Egyptian and Mesopotamian medicine and pharmacy had a profound impact on shaping the future of Western medicine and pharmacy. The ancient Hebrews who lived amid both of these ancient civilizations observed Egyptian and Mesopotamian practices and adapted them to fit within the monotheistic framework which would become their living legacy to Western civilization. The ancient Greeks and Romans too were profoundly affected by Egyptian and Mesopotamian medical and pharmacy practices.

Chapter Summary

This chapter describes how the Mesopotamians developed a system of explaining illness and treating it using an amazing array of plants, minerals, and animal products. They also created a full line of cosmetics for both men and women.

Due to their geographic isolation, the Egyptians enjoyed long periods of peace and used this stability to establish a division of labor and a degree of specialization in pharmacy that was remarkable for this era. They pioneered the use of many herbs, such as castor beans and senna which are still used today in many laxatives. The Egyptians also advanced the use of cosmetics.

Key Terms

Sumerians	Hammurabi's Code	*wekhudu*
cuneiform	Thoth	hieroglyphics
influenza	Anubis	Rosetta Stone
caduceus	Isis	Jean François
baru	Sekmet	Champollion
ashipu	Imhotep	Ebers papyrus
asu	Edwin Smith papyrus	*asi-t; pastophors*
dosage form	*metu*	

Chapter in Review

1   Describe the Mesopotamian theory of illness.

2   Explain how the Mesopotamians diagnosed illness and treated it.

3   Identify the roles played by the *baru*, *ashipu*, and *asu*.

4   Describe the dosage forms developed by the Mesopotamians.

5   Describe the Egyptian theory of illness.

6   Explain how Egyptian priest-physicians diagnosed and treated patients.

7   Describe the Egyptian medical system and how pharmacy fit into it.

8   Identify the unique characteristics of Egyptian medicine and pharmacy.

Notes

1 Otto A. Wall, *The Prescription Therapeutically, Grammatically and Historically Considered* (St. Louis, MO: August Gast Bank-Note and Litho. Company), 179.

2 Sigmund Freud, *Civilization and Its Discontents*, ed. and trans. James Strachey (New York: Norton and Company, 1961), 59.

3 Albert S. Lyons and R. Joseph Petrucelli, II, *Medicine: An Illustrated History* (New York: Abradale Press, 1987), 63. H.W.F. Saggs, *Civilization Before Greece and Rome* (New Haven, CT: Yale University Press, 1989), 262.

4 James Grier, *A History of Pharmacy* (London: The Pharmaceutical Press, 1937), 15–16.

5 David L. Cowen and William H. Helfand, *Pharmacy: An Illustrated History* (New York: Harry Abrams, 1990), 19.

6 Charles Lawall, *The Curious Lore of Drugs and Medicine: Four Thousand Years of Pharmacy* (Garden City, New York: Garden City Publishing, Inc., 1927), 17.

7 Roberto Margotta, *The Story of Medicine* (New York: Golden Press, 1968), 21.

8 James Grier, *A History of Pharmacy*, 213.

9 Albert S. Lyons and R. Joseph Petrucelli, II, *Medicine: An Illustrated History*, 67.

10 Roberto Margotta, *The Story of Medicine*, 22.

11 Samuel Noah Kramer, *The Sumerians: Their History, Culture, and Character* (London: The University of Chicago Press, 1963), 95.

12 Glenn Sonnedecker, Compiler, *Kremers and Urdang's History of Pharmacy*, (4th ed., Madison, WI: American Institute of Pharmacy, 1976), 5.

13 David L. Cowen and William H. Helfand, *Pharmacy*, 20.

14 C.J.S. Thompson, *The Mystery and Art of the Apothecary* (London: John Lane the Bodley Head, Ltd., 1929), 6.

15 James Grier, *A History of Pharmacy*, 13.

16 Jenny Sutcliffe and Nancy Duin, *A History of Medicine* (London: Barnes and Noble Books, 1992), 14.

17 George Bender, *Great Moments in Pharmacy* (Detroit, MI: Northwood Institute Press, 1966), 14.

18 Samuel Noah Kramer, *The Sumerians*, 97.

19 James Grier, *A History of Pharmacy*, 12.

20 Jenny Sutcliffe and Nancy Duin, *A History of Medicine*, 14.

21 Joseph B. Sprowls, Jr. and Harold M. Beal, eds., *American Pharmacy: An Introduction to Pharmaceutical Technics and Dosage Forms* (6th ed., Philadelphia: J.B. Lippincott Company, 1966), 4.

22 C.J.S. Thompson, *The Mystery and Art of the Apothecary*, 10.

23 Sally Pointer, *The Artifice of Beauty* (Gloucestershire: Sutton Publishing, Ltd., 2005), 11.

24 Ibid., 15.

25 James Grier, *A History of Pharmacy*, 6. C.J.S. Thompson, *The Mystery and Art of the Apothecary*, 11.

26 Anne Rooney, *The Story of Medicine* (London: Arcturus Publishing, Ltd., 2009), 172.

27 Diarmuid Jeffreys, *Aspirin: The Remarkable Story of a Wonder Drug* (London: Bloomsbury, 2004), 9.

28 A.C. Wootton, *Chronicles of Pharmacy, Vol. 2* (Reprint: Boston, MA: Milford House, Inc., 1972), 23–24. Mary Roach, *Stiff: The Curious Life of Human Cadavers* (New York: Norton, 2003), 222–225. C.J.S. Thompson, *The Mystery and Art of the Apothecary*, 212–213.

29 David L. Cowen and William H. Helfand, *Pharmacy*, 22. Glenn Sonnedecker, *Kremers and Urdang's History of Pharmacy*, 9.

30 Charles Lawall, *The Curious Lore of Drugs and Medicine*, 5–9. C.J.S. Thompson, *The Mystery and Art of the Apothecary*, 13. A.C. Wootton, *Chronicles of Pharmacy, Vol. 1* (1910, Reprint: Tuckahoe, New York: Milford House, Inc., 1972), 43.

31 George Bender, *Great Moments in Pharmacy*, 20. Glenn Sonnendecker, *Kremers and Urdang's History of Pharmacy*, 8–9.

32 Joseph B. Sprowls, Jr. and Harold M. Beal, eds., *American Pharmacy*, 5.

33 Charles Lawall, *The Curious Lore of Drugs and Medicine*, 10.

34 George Bender, *Great Moments in Pharmacy*, 22. David L. Cowen and William H. Helfand, *Pharmacy*, 23.

35 Sally Pointer, *The Artifice of Beauty*, 19.

36 Ibid., 19.

37 "Ancient Invention of War, Sex, and City Life," dir. Daniel Percival, prod. Tomi Bednar Landis, Discovery Channel, 2008. Glenn Sonnedecker, *Kremers and Urdang's History of Pharmacy*, 7. Albert S. Lyons and R. Joseph Petrucelli, II, *Medicine*, 90.

38 Sally Pointer, *The Artifice of Beauty*, 21–22.

39 Ibid., 20.

40 Albert S. Lyons and R. Joseph Petrucelli, II, *Medicine*, 90.

41 Ibid., 97. Roy Porter, *The Greatest Benefit to Mankind: A Medical History of Humanity* (New York: Norton Books, 1997), 49.

42 Jenny Sutcliffe and Nancy Duin, *A History of Medicine*, 12.

3 Pharmacy and Medicine in the River Civilizations of the East

Ancient India and China

The physician, drug, nurse, and patient are the four pillars upon which is supported the treatment of diseases.

(Kaviratna)[1]

Ancient India: Ayurvedic Pharmacy and Medicine

How did Ayurvedic medicine start? How does it work? How has it been able to survive over the millennia?

Between 2500 and 1500 BCE three cities emerged from their Neolithic shadows on the Indian subcontinent near the Indus River Valley—Mohenjo-Daro, Harappa, and Lothal. The cities housed tens of thousands of people and featured public water and sewer systems including wells, latrines, and public baths. From the start, public health was a concern where ritual bathing for religious reasons made for good hygiene. The cities had well planned grids of streets replete with sturdy housing made of burnt brick. Around 1500 BCE, people from Central Asia known as the Aryans conquered this early civilization and exercised a profound influence over its development.[2]

One of the world's great religions, Hinduism, emerged during this era and would shape the development of ancient Indian pharmacy and medicine through the precepts of its scripture known as the Vedas (Sanskrit for "knowledge"). The four Vedas include the *Rig-Veda*, the *Yajur-Veda*, the *Sama-Veda*, and the *Atharva-Veda*. According to the *Rig-Veda*, disease was a function of bad karma or personal fate in which a person was punished by divine wrath for bad behavior.[3] According to the creation story, Dhanvantari became the god of medicine and passed down the secrets of life to humanity in the form of medical texts. Although the *Atharva-Veda* was largely liturgical, it contained prayers and magical spells used to treat patients suffering from illness or injury. Still, the *Atharva-Veda* offered a prescient description of a deadly disease that begins with an eruptive fever accompanied by spots on the body, in a word smallpox.[4] There was an additional Veda known as the *Ayur-Veda* (or knowledge for longevity) from which two key medical treatises were derived, known as the *Caraka* and *Susruta-Samhitas*.

Chart of Hindu Medical Gods

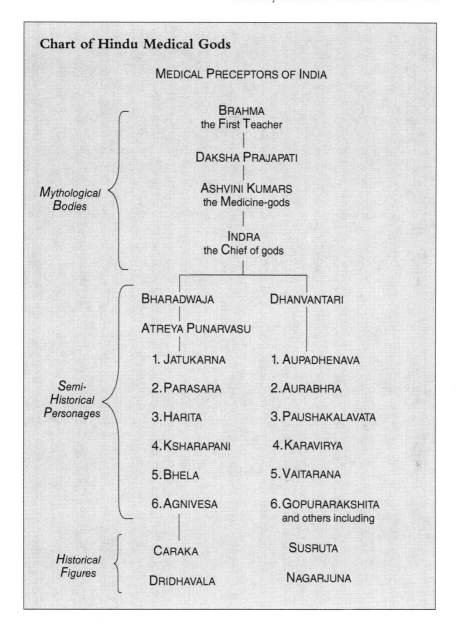

MEDICAL PRECEPTORS OF INDIA

Mythological Bodies
- BRAHMA
  the First Teacher
- DAKSHA PRAJAPATI
- ASHVINI KUMARS
  the Medicine-gods
- INDRA
  the Chief of gods

Semi-Historical Personages

BHARADWAJA

ATREYA PUNARVASU

1. JATUKARNA
2. PARASARA
3. HARITA
4. KSHARAPANI
5. BHELA
6. AGNIVESA

DHANVANTARI

1. AUPADHENAVA
2. AURABHRA
3. PAUSHAKALAVATA
4. KARAVIRYA
5. VAITARANA
6. GOPURARAKSHITA
   and others including

Historical Figures

CARAKA

DRIDHAVALA

SUSRUTA

NAGARJUNA

The Birth of Ayurveda

Recent scholarship has suggested that the *Ayur-Veda* came about as the result of ascetic communities that emerged in northern India. One of the best known of these communities was established by Siddhartha Gautama Sakyamuni (*c.*563–483 BCE), better known as the Buddha. Soon, Buddhist monks formed monastic communities to meditate and to follow a strict form of moral discipline. Leading a simple life, these monks identified five basic medicines including fresh butter, clarified butter (*ghee*), oil, honey, and molasses. Over time, the list of medicines grew.[5] Buddhist monasteries often maintained "sick rooms" that would later provide the impetus to build hospitals. King Asoka (269–237 BCE), one of India's greatest monarchs of the Mauryan dynasty, converted to Buddhism after having conquered much of India, and established hospitals. Thus, ancient India became, perhaps, the earliest civilization to embark upon a path of state-sponsored medicine, with hospitals playing a prominent role.[6]

In fact, the *Caraka-Samhita* described the ideal hospital in detail:

> The engineer is to erect a strong and spacious building, well-ventilated at one part, the other part being free from draughts. The scenery should be pleasing and one should feel happy to walk in it. It must not be behind any high building, nor exposed to the glare of the sun. It should be inaccessible to smoke and dust. There must not be anything injurious to our senses as regards sound, touch, taste, form, and smell. There should be stairs, large wooden mortars and pestles; and there must be additional bare ground for the construction of a privy, bath-room, and kitchen.[7]

In similar terms, the *Caraka-Samhita* described the ideal hospital staff:

> The staff should consist of servants and companions. The servant should be good, virtuous, pure, fond, clever, generous, well trained in nursing, skillful in work . . . practiced in the art of compounding medicines and a willing worker not likely to show displeasure to any order.[8]

Ayurvedic medicine and its great medical texts were influenced by both the Vedic and Buddhist traditions. The *Caraka-Samhita* or *Caraka-Compendium* dealt with medicine and pharmacy, while the *Susruta-Samhita* dealt with surgery, which for its time, was one of the most advanced systems in the world. The dates of when these treatises were written range from 1000 BCE to 1000 CE. There is a weak consensus among experts that the *Caraka-Samhita* dates back as far as the first century CE followed by the *Susruta-Samhita* in the fourth century CE.[9]

The *Caraka* and *Susruta-Samhitas* described the appearance and medicinal use of more than 2,000 substances including aconite, cardamom, cinnamon, ginger, licorice, pepper, and sandalwood. Some of the more remarkable discoveries include snakeroot (*Rauvolfa serpentina*) for sedation, deadly nightshade (*Atropa belladonna*), Indian hemp (*Cannabis indica*) to induce stupor, and chaulmoogra oil to treat leprosy. *Rauvolfa serpentina* is the source of reserpine, which is widely used in modern anti-hypertension and anti-depression drugs.[10]

The Caraka-Samhita

The *Caraka-Samhita* lists some 177 substances derived from animals that have therapeutic value in treating over 200 diseases and 150 other conditions. Animal excrement, urine, flesh, fat, blood and milk from animals ranging from snakes to horses were mentioned. Based on Hindu religious belief, products from cows were especially viewed to have purifying properties.[11] Similarly, a variety of minerals *Parthiva* (lifeless) were mentioned including gold, silver, copper, lead, tin, iron as well as sand, lime, and various salts.[12] All of these drugs were classified by the diseases they were supposed to treat.

Dosage Forms

Ancient Ayurvedic pharmacy included the use of many dosage forms that were mindful of the power of dosages when administering them. Susruta himself instructed that the dose of a drug should be prepared in accordance with the age of the patient. Special care should be taken in case of old age or in the case of children.[13] Medicines were prepared and administered in the form of poultices, pastes *(lepa)*, fumigations, inhalations, suppositories, infusions, decoctions, lotions, and oral forms.[14]

Caraka's Code of Ethics

The *Caraka-Samhita* emphasized the importance of ethical medical practice in its Oath of Initiation which is strikingly similar to the Hippocratic Oath. The *Caraka-Samhita* called upon its initiates to repay their teachers, to be free of envy, and never carry weapons. It also stated:

> Every day you should continuously and whole-heartedly try to promote the health of patients. Even if your own life is in danger you should not neglect the patient . . . You should not entertain an evil thought about the wealth or wives of others. . . . You must treat as strictly confidential all information about the patient and his family.[15]

These were the characteristics of the good physician *(vaidya)*.

The Mysterious Wonder Drug: Soma

During ancient times one of the most powerful drugs used was soma, a narcotic used to relieve pain. The drug, which some experts believe derives from the hallucinogenic mushroom *Amanita muscaria*, became so widely used and abused that the authorities banned its use. Today, several theories abound about what the actual ingredients of soma were and whether it still exists in the form used by the ancients.[16] After 500 CE, alchemy with its emphasis on metals, especially mercury, played a key role in advancing Ayurvedic pharmacy. A form of medicine based on this tantric-alchemical era during the medieval era that is still practiced in the southern Tamil-speaking areas of India is known as Siddha medicine based on the Tamil word *cittar*.[17] In the eighth and ninth centuries, India came into greater contact with Persia and then the Arab world. Early Islamic caliphs such as al-Mansur and Harun al-Raschid established contacts with India and encouraged their envoys to learn as much as they could about mathematics, science, and medicine. During the eleventh century, Islamic invaders from Central Asia conquered India and brought their form of medicine with them known as *Unani* (or Ionian meaning Greek). Unani medicine was the Western Galenic medicine as refined and advanced by Avicenna (980–1037). Despite centuries of Western colonial influence after the sixteenth century CE, Ayurvedic pharmacy and medicine survived.[18]

Ayurvedic Medical Theory

According to Ayurvedic medicine, there are six fundamental bodily elements including chyle, blood, flesh, bone, marrow, and semen. In addition to the six bodily elements, each of us is born with a unique ratio of three life forces or humors known as *doshas*. The *doshas* include *vata*, *kapha*, and *pitta*. *Vata* is symbolized by air and represents the bodily energy that causes humans to move. *Kapha* represents the earth (phlegm) and is symbolized in the body as structure and includes bones, muscles, and teeth or bodily structure. *Pitta* represents fire (bile) and is symbolized in the body as digestion or the transformative force between *kapha* and *vata* that provides the energy for us to move. These *doshas* circulate within the body through 13 bodily channels known as *srotas*. Each *dosha* has *gunas* or fundamental qualities. For example, *vata*'s fundamental qualities are dry and cold. So, how does this work? If someone eats a bowl of hot chili, *pitta* increases in the body, and to balance hotness the person drinks cool water to retain the body's state of equilibrium. Each individual has a unique mix of *doshas* and the Ayurvedic physician in consultation with the patient needs to determine the patient's dominant body type based on *dosha*. For example, a person whose body is dominated by *pitta* tends to be passionate, argumentative, and quick to anger. *Pitta* body types tend to suffer from fiery ailments such as rashes, acid reflux, and peptic ulcers.[19]

Susruta: The First Plastic Surgeon

One of the outstanding figures in Ayurvedic medicine was the gifted surgeon Susruta who lived sometime during the fourth century CE. The *Susruta-Samhita* which was attributed to him, established the ethics, education, and standards of surgical performance that made Ayurvedic surgeons the most advanced of their time. Using a wide variety of over 100 surgical instruments

Figure 3.1 Susruta was a legendary surgeon of ancient India who pioneered plastic surgery more than 2,000 years ago.

Source: Painting by Robert Thom. Collection of the University of Michigan Health System, Gift of Pfizer Inc. UMHS.6.

that modern surgeons still use, Ayurvedic surgeons performed a number of procedures including incision, excision, scraping, puncturing, extraction, secreting fluids, and suturing. Ayurvedic surgeons practiced surgical techniques on melons and vegetables before practicing on patients. Even though their knowledge of anatomy was far from complete they identified 107 *marmas* or vital points on the body that surgeons needed to avoid when performing surgery lest they harm patients irreparably.

They even invented biodegradable sutures by applying the heads of Bengali fire ants to close wounds. By patiently holding a fire ant over the wound, the surgeon waited for the ant to clamp its jaws on the wound, and then the surgeon would cut off the ant's thorax, and would repeat the process as needed to close the wound.[20] If that failed, surgeons also employed cautery especially when performing amputations. Leeches were sometimes used to draw blood from contusions and lesions to relieve swelling. A wide array of bandages were used in surgery. Wine laced with hemp and other narcotics was used as anesthesia. Hypnosis also was practiced. Perhaps the most remarkable surgical procedures include rhinoplasty and otoplasty.

Ayurvedic surgeons performed plastic surgery to replace severed noses and ears by taking skin from the patient's cheek or forehead and then attaching it to fashion a new nose or ear for the patients. These procedures would not be attempted in the West until the sixteenth century.[21]

Medicinal Leeches

Leeches have been long associated with the art of medicine and since the 1980s have made an astounding comeback in promoting healing after complex microsurgeries. Introduced in ancient India's Bengal region, leeches were used as a treatment to rid the patient of "bad blood." In ancient Greece, Themison (*c.*50 BCE) made note of them in the first century BCE followed by the Roman encyclopedist Pliny in the first century CE. From this point forward, the leech had become a mainstay of medical treatment as evidenced by the medieval Anglo-Saxons in England who actually referred to their barber-surgeons as "leeches." Medieval medical books were entitled "leechbooks."

Since colonial times, nearly every American apothecary shop had a leech jar prominently displayed on its counter. Leeches fell out of favor with the rise of germ theory in the late nineteenth century but made a stunning comeback in the 1980s when they were used in the treatment of contusions and wound healing. Modern researchers have found that when a leech attaches itself to a human, it injects a natural anesthetic and feasts on impurities in the host's blood. The leech stays attached until it engorges itself to about six times its original size and then sloughs off, contently fed. Today, the world's leading leech farms are located in Wales and Finland and medicinal leeches are stocked in nearly every hospital that performs grafting or microsurgeries.

Leeches are especially effective in promoting healing after delicate microsurgeries such as the reattachment of severed body parts such as fingers and toes. Modern surgeons are skilled at reattaching larger blood vessels such as arteries and veins, but not small capillaries. The medicinal leech attaches itself to the recently reattached finger and, by sucking on the blood, keeps the incoming blood from pooling, which could otherwise cause complications. The leech keeps the blood flowing until the capillaries and veins can re-establish themselves.[22]

Susruta on Building Dispensaries and Storing Medicines

Although a surgeon, Susruta offered advice on constructing dispensaries in a clean area with the building facing a cardinal direction either east or north. He suggested that medicines should be kept in burnt earthen pots arranged on planks supported by stakes or pins. Susruta also advised,

The wise physician should collect and classify these medicines and with them prepare external applications, infusions, oils, ghee, syrups, etc., as required for the derangement of a particular *dosha*. The medicines should be carefully preserved in all seasons, in rooms free from smoke, rain, wind, and dust. The medicines should be used singly, or in combinations of several medicines of a class, or of an entire group, or of more than one group, according to the nature of the diseases, and the extent of derangement of the *doshas*.[23]

Ancient China

How did Chinese medicine start? What is its theory of illness?
What is the secret of its longevity as a medical system?

Around 3000 BCE, people from Central Asia settled in the Yellow River valley and established one of the world's great civilizations. By 1100 BCE, the civilization had expanded south to the Yangtse River valley. Although Chinese civilization would make many cultural contributions to humanity, their interest in advancing pharmacy and medicine remains among their most outstanding and timeless. Being a great river civilization, Chinese pharmacy and medicine were greatly influenced and highly circumscribed by their emperors and educated officials known as mandarins. The earliest of these was the legendary Emperor Fu Hsi (*c.*2900 BCE, who supposedly created the symbol of the *pa kua*, which symbolized the *tao* or way of the universe.[24] According to the *pa kua*, the universe is divided into two complementary forces that are in flux. The *yin* connotes the feminine, passive, cold, wet, and dark principle of nature while the *yang* connotes the masculine, active, warm, dry, and light principle of nature. Just as the *pa kua* symbolizes the key forces of the universe on a macro level, so too does it apply on the micro level to the human body.[25]

Worm Spirits

In its early stage, Chinese medicine was quite mystical. Diseases were believed to be caused by demons in the form of worm spirits or *ku*. Oral rites were administered that included incantations, dance, drugs, and centipedes, because centipedes fed on worms. *Dreckapotheke* or strong drugs were usually administered as fumigations or amulets were hung around the neck of patients to drive out the demons and keep them away. In some cases, patients resorted to sorcerers to magically transfer the worm spirits to other beings for relief. According to legal documents, *ku* magic persisted well into the nineteenth century. Penalties for engaging in *ku* magic were quite harsh and included bizarre capital punishment for those convicted.[26]

The Five Basic Elements

The ancient Chinese discovered five basic elements of nature including wood, fire, earth, metal, and sky, and named how these elements interact through transitional phases of flux called *wu xing*.[27] In addition to the five basic elements of nature, the ancient Chinese identified five planets, five seasons, five directions, five colors, five sounds, and five human organs. They called illness *"bing"* and defined it as a state of disharmony among the five organs, a disharmony connected to the disharmony of the planets, seasons, colors, and sounds corresponding to each organ. For them, all conditions were complexes. During an examination of a patient the Chinese physician took the patient's pulse and inspected the color of the patient's tongue. Chinese physicians identified 51 types of pulses and 37 shades of the tongue. Chinese physicians identified a number of diseases including diabetes, dysentery, measles, cholera, smallpox, and bubonic plague. To protect patients from contracting smallpox, the Chinese practiced

> Dragon's teeth or dragon's bone was a widely used ingredient in many early Chinese prescriptions. Dragon's teeth were ground up bits of prehistoric fossils that were used to treat conditions ranging from lung ailments to anxiety.[28]

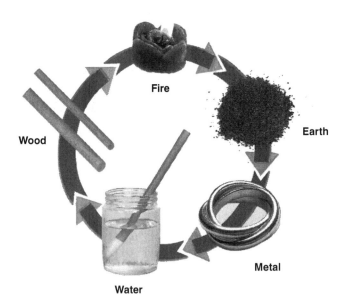

Fire

Wood

Earth

Metal

Water

Figure 3.2 The Five Elements of Chinese Medicine or *wu xing* represent the Taoist forces of yin and yang and interact in a metaphysical way to account for all substances in the universe.

variolation. They would remove the scabs from a patient infected with small-pox, grind them into a powder, and then have patients inhale the powder. They also closely monitored rat mortality. If rats died in large numbers, they would abandon towns and move to safer locations in order to avoid bubonic plague.[29]

Shen-nung: The Red Emperor and Sage of Agriculture and Pharmacy

The Chinese were optimistic about treating illness which was reflected in their assumptions about health and illness. They believed that for every ill a remedy existed and that everything in nature had a potential for medical use. These assumptions led the Chinese to place a great emphasis on drug therapy to cure illness. Their great legendary emperor Shen-nung acted as a one-man herb testing laboratory, personally testing thousands of substances for potential medical use. Shen-nung also known as the Red Emperor was said to have lived about 2500 BCE; it was more likely around 400 BCE. Shen-nung was

Figure 3.3 Shen-nung was a legendary emperor of ancient China who pioneered the use of herbal medicine, a mainstay of Chinese medicine today. He is credited with the discovery of tea.

Source: Painting by Robert Thom, reproduced courtesy of the American Pharmacists Association Foundation.

credited with writing an herbal book known as the *Pen-Ts'ao*, which listed 365 medicinal substances, prescriptions (*fang*), and poisons. The *Pen-Ts'ao* was an oral tradition that had been passed down through generations and was eventually recorded by scholars over the centuries. The key substances described included camphor, cassia bark, ephedra (*ma huang*), ginseng, kaolin, opium, and rhubarb. Ephedra was used to treat lung disorders, kaolin was used to treat diarrhea, opium was used as a narcotic, and rhubarb was used as a laxative.[30]

According to Chinese folklore, Shen-nung discovered tea one day when a tea leaf serendipitously fell into a cauldron of water he was boiling. Similarly, he showed the Chinese how to farm using a plow, which accounts for why he was sometimes depicted in portraits as having ox horns on his head. Shen-nung's work was expanded over time and most significantly enhanced by the work of Li Shizhen (1518–1593), whose *Pen-Ts'ao Kang-mu* (*Great Herbal*), was published in 1596. The *Pen-Ts'ao Kang-mu* consists of 52 volumes that describes over 1,000 plants, 450 animal substances, and 10,000 remedies for various ailments with 1,000 illustrations. Each drug came with a booklet describing the drug, its locale, methods of procurement, its odor, its taste, principal application, its preparation, and an appraisal of its medicinal value. Some of the more intriguing uses of animal substances include pulverized sea horses (that contain natural iodine) to treat goiter, snake meat to treat eye disorders, octopus ink mixed with vinegar to treat heart ailments, and elephant skin to treat sores. Herbal drug therapy was, and continues to be an important part of Chinese medicine today.

Chinese Medical Specialization

During the Chou Dynasty (*c.*1122–255 BCE) the government regulated the education and practice of physicians through the Imperial College of Medicine which conducted examinations of aspiring physicians. Those who earned the highest scores were admitted into the Imperial Service and their performance was monitored in terms of how many patients they cured. The ethic for the Chinese physician was that good doctors kept their patients healthy, while bad doctors treated sick patients.[31] The Imperial Service also established four specialties for physicians that included doctors of dietetics (*shi yi*) (apothecaries), doctors of internal diseases (*ji yi*), doctors of sores (*yang yi*) (trauma), doctors of veterinary medicine (*shou yi*), and doctors who supervised the others.[32] By 318 BCE, there was mention of a pharmacy at the Imperial Court.[33]

Sun Szu-miao

Since Shen-nung's time there were several *shi yi* (physician/apothecaries) who made major contributions to the advancement of Chinese pharmacy in both its herbal and alchemical traditions. Sun Szu-miao (*c.*581–682) wrote *Ch'ien Chin Yao Fang* (*Vital Prescriptions Worth a Thousand Gold Pieces*), a wide-ranging 30-volume herbal which, among other things, deals with the collection, preservation, preparation, and administration of drugs; 863 of them. One of the remedies, *Dichroae (chagshan)*, had been recently introduced from Central Asia as a remedy for malaria. A large part of the work dealt with preventive medicine that provided advice about nutrition, sanitation, massage, and exercise. Sun Szu-miao's work synthesized the Taoist mystical penchant for folk remedies with Confucian rationalism, and drew heavily on both traditions. His work also dealt with medical treatments and anticipated the further specialization of physicians that were put into practice by the end of his life.[34]

Lu Yu

Lu Yu (733–804) was the rare combination of poet and man of science who wrote *The Classic of Tea*. The work was written in three books consisting of ten chapters on every aspect of cultivating, brewing, and drinking tea. One day as he was searching for new varieties of tea, he stumbled upon a spring of pure water. He brewed some tea in it, and realized that the key to a good cup of tea resided in the quality of the water in which it was brewed. He also surmised that drinking clean and pure water was not only the key to a good cup of tea, but was the key to good health and an important preventive measure against disease.[35] Of course, today, media reports galore exhort the health benefits of drinking green tea, and from a public health standpoint, the health benefits of drinking clean water are indisputable.

Despite the pressures to conform to Neo-Confucian standards during the twelfth and thirteenth centuries, physicians/scholars such as Tang Sheweni continued to advance Chinese pharmacy. During the Sung dynasty, Tang Sheweni wrote the *Zhenglei Bencao (Systematic Pharmacology)* in 1159 that contained 1,740 remedies. Similarly, Zhang Yuansu (*c*.1151–1234), another physician/scholar, wrote the *Zhenzhu Nang (Bag of Pearls)* that reconciled Chinese drug therapy with its medical philosophy and theory. For example, he integrated and classified herbs according to the five elements (*wu xing*). Zhang Yuansu's work maintained that herbs had unique tastes that affected different organ systems and that they entered into meridians affecting them as well. His work showed that the Chinese regarded food as medicine.[36]

Li Shizhen

Li Shizhen (1518–1593) was a polymath, a Renaissance man in China. Physician, apothecary, and botanist, Li Shizhen spent 27 years researching and writing China's *Pen Ts-ao Kang-mu (The Great Materia Medica)* that was published posthumously in 1596. In 52 chapters, he described over 1,892 remedies, and based on his discoveries, 374 of them for the first time. Moreover, the work contained 10,000 prescriptions and 1,000 illustrations. Some of the more advanced prescriptions included the use of iodine, mercury, and ephedrine to treat goiter, venereal disease, and lung ailments, respectively. Li Shizhen followed in his father's footsteps and became a physician. He cured a prince and became a government official, a development that afforded him access to all of the medical books that existed at the time; numbering about 800. Confused by the contradictions he read in them, he set his life's course in writing a definitive book that would correct the errors and add the knowledge he gained through his own extensive travels and experience. Two centuries after Li Shizhen's work was published it was followed by an expanded edition written by Zhao Xuemin that described 2,608 different remedies, a total that has not been surpassed.[37]

Chinese Alchemy

Similar to ancient India, apothecary/physicians and others became interested in alchemy. While China had its share of alchemists who were obsessed with turning mundane substances into gold, there were those who searched for new drugs to cure human illness. Chinese alchemy followed two main traditions: *waidan* (external alchemy) and *neidan* (internal alchemy). Ko Hung (*c.*300) experimented with elixirs that could aid digestion or ward off wild animals and ghosts. Similar to what would happen in the West later, Chinese alchemists sought the elixir of immortality and in that search stumbled upon some useful medicinal remedies.[38]

Although acupuncture had been known to the West since the seventeenth century, it gained widespread attention in the West in 1971 as the result of an American journalist's case of appendicitis. James Reston, a columnist for the *New York Times*, was part of the press corps that accompanied President Richard Nixon during his historic trip to the People's Republic of China in July 1971. Reston suffered a case of acute appendicitis that needed immediate medical attention. There was no time for medical evacuation, so at the behest of Premier Chou Enlai, Reston had his appendix removed at Beijing's Anti-Imperialist Hospital on July 17, 1971. Reston had been given a local anesthetic, and remained conscious during the procedure. Afterward he received traditional acupuncture and moxibustion to treat post-operative discomfort. Reston reported on his experience in his column and Chinese medicine, especially acupuncture, gained renewed interest in the West.[39]

Huang-ti: The Yellow Emperor

The Yellow Emperor Huang-ti, the third of the legendary emperors, lived around 2600 BCE and was primarily known for unifying China and supervising the construction of the Great Wall. During his reign the *Nei Ching* (*Canon of Medicine*) was written and forms the basis of Chinese medicine. Scholars place the actual writing of the *Nei Ching* at about 200 BCE. The methods of treatment described include acupuncture, minor surgery, drugs, prayer, and moxibustion. The standard Chinese treatment was called *zhenjiu* or a combination of "needles and moxa." The best known treatment described was acupuncture which the Chinese have been practicing for over 2,500 years. The treatment consists of inserting very thin needles (Latin *acus*) into key impulse points on the body (called *fornina*) to keep *qi* or energy flowing along the body's meridians where they are then twirled and vibrated. *Qi* makes a complete circuit around a healthy person's body once every two hours. The goal of acupuncture is to restore this natural flow of energy and restore the body's natural balance. Another variant of acupuncture treatment was moxibustion where the powdered cones of the leaves of mugwort or wormwood plants are burnt into the skin or placed at the end of acupuncture needles to restore the flow of *qi* throughout the body. Acupuncture came to the West in the seventeenth century but has become much more popular in the United States since the 1970s. Today, the World Health Organization has named 40 conditions that benefit from acupuncture therapy.

Bian Que and the Nanjing (Classic of Difficult Cases)

The first actual physician mentioned in the imperial history was Bien Que, who lived around 200 BCE. Bien Que, by all accounts, was a talented diagnostician whose gift relied on his uncanny ability to read a patient's pulse. One of the works attributed to him was on reading pulses, but it has not survived. The other medical treatise that has survived was the *Nanjing* or *Classic of Difficult Cases* that scholars think Bien Que may have started, but was expanded by generations of his students. Tragically, Bien Que was killed by an assassin hired by a jealous rival, Li Xi, who was the master of the guild of physicians in Qin province.[40]

Zhang Zhongjing

Zhang Zhongjing (*c*.150–219) grew up amid civil war and pestilence and became a physician. He wrote the *Shanghan Zabinglun* or (*Treatise on Cold Pathogenic and Miscellaneous Diseases*) which was a compendium of 22 separate monographs that contain about 400 pieces of practical advice on treating various diseases. He was an early advocate for the use of steam baths, hydrotherapy, and enemas. Much of the original *Shanghan Zabinglun* was lost, but what remains of it can be attributed to the work of another physician, Wang Shuhe (*c*.265–317), who copied the first six books during the 280s.

Hua Tuo

Hua Tuo (*c*.140–208) was a contemporary of Zhang Zhongjing. Similar to his contemporary, Hua Tuo also believed in the value of hydrotherapy and exercise regimens that might have been a precursor to *tai chi*. What sets Hua Tuo apart from other Chinese physicians was his skill as a surgeon and the fact that he grew up in Central Asia and had been influenced by the Buddhist

communities there. He used a powerful anesthesia called *mafiesan* which pur-
portedly was made by boiling cannabis powder and dissolving it in wine. As
a successful surgeon he also must have developed some type of anti-infection
drug(s), but similar to the formula for *mafiesan*, this knowledge was lost.[41]

Hua Tuo's renown as surgeon reached its pinnacle in an incident in which
he successfully removed an arrow from the shoulder of a warlord, the Duke
of Kuan. The duke fully recovered and demand for Hua Tuo's skill as a surgeon grew. The most powerful official in the Three Kingdoms, Cao Cao,
made Hua Tuo his personal physician. Hua Tuo, who was a scholar as well
as a physician, became bored and homesick. He left the court and returned
home to his wife and family, whereupon Cao Cao ordered him to return.
When Hua Tuo dallied, Cao Cao ordered him to be brought back and executed. Several officials tried to intercede on Hua Tuo's behalf, but to no avail.
While awaiting his execution, Hua Tuo reportedly wrote a scroll containing
the knowledge of his life's work and offered it to the prison warden for the
benefit of posterity. The warden, fearing reprisals by accepting a scroll from a
condemned man, refused the scroll, whereupon Hua Tuo burnt it.[42]

Hua Tuo's mercurial rise and fall as a Chinese court surgeon represented
an instructive and cautionary tale about the dangers of introducing innovative science in a Confucian system that had little tolerance for free thinking.
With the exception of the procedure of castration, Hua Tuo's other surgical
procedures were lost. After Hua Tuo surgery became a backwater in Chinese
medicine and its revival would not occur until the twentieth century.

The Zhubing Yuanhoulin *or* (Notes on the Origins and Courses of All Diseases)

For several centuries the most definitive medical encyclopedia used in China
was compiled by a group of physicians under the supervision of the scholar
Chao Yuanfang. The *Zhubing Yuanhoulin* was a wide-ranging work, composed of 50 chapters describing the diagnosis and prognosis of 1,720 different
disorders. The diseases described include smallpox, bubonic plague, dysentery, and measles among others. The encyclopedia, which was published and
distributed by the imperial government in 610, also provided the first description of beriberi (a vitamin deficiency of the B complex). The encyclopedia
represents the best window we have into the state of Chinese science and
medicine during that time.

Chapter Summary

Ayurvedic pharmacy and medicine has had a dramatic impact on the development of pharmacy and medicine in Asia similar to the influence that ancient
Greek medicine had in the development of Western pharmacy and medicine.

As we shall see in upcoming chapters, Ayurvedic medicine and ancient Greek medicine share much in common and historians continue to speculate about how these two systems have mutually influenced each other. Although still viewed in the West as a form of alternative medicine, Ayurvedic medicine continues to make inroads in the West as people the world over still search for health and well-being.

Similar to Ayurvedic medicine, Chinese medicine remains timeless and competes with Western medicine. From its outset, with the legendary emperor Shen-nung, the Chinese have attached great importance to drug therapy to treat illness and to this day the modern Chinese pharmacy carries a vast array of herbs and drugs that is unrivalled in the world.[43]

Key Terms

Mohenjo-Daro	*gunas*	Li Shizhen
Sanskrit	Susruta	*Pen-Ts'ao Kang-mu*
Ayur-Veda	*marmas*	*shi yi*
Dhanvantari	Fu Hsi	Sun Szu-miao
ghee	*pa kua*	Lu Yu
King Asoka	*tao*	*waidan*
Caraka-Samhita	*yin*	*neidan*
Susruta-Samhita	*yang*	Huang-ti
soma	*ku*	*Nei Ching*
Siddha medicine	magic	*qi*
Unani medicine	*wu xing*	acupuncture
dosha	*bing*	moxibustion
vata	Shen-nung	Bian Que
kapha	*Pen-Ts'ao*	Nanjing
pitta	*fang*	Zhang Zhongjing
srotas	*ma huang*	Hua Tuo

Chapter in Review

1 Describe the origins of ancient Indian civilization and medicine.
2 Explain the Ayurvedic theory of illness.
3 Explain how Hinduism and Buddhism affected the development of Ayur-vedic medicine.
4 Describe the contents of the *Caraka-Samhita* and *Susruta-Samhita*.

5   Describe the Chinese theory of illness.
6   Explain how Chinese physicians diagnosed and treated illness.
7   Explain the philosophical system that guides Chinese medicine.
8   Trace the development of Chinese pharmacy from its origins.
9   Describe China's contribution to pharmacy and medicine.

Notes

1   G.P. Srivastava, *History of Indian Pharmacy, Vol. 1* (Calcutta: Pindars, Ltd., 1954), 125.
    M. Patricia Donahue, *Nursing: The Finest Art* (2nd ed., St. Louis, MO: Mosby, 1996), 47.
2   Albert S. Lyons and R. Joseph Petrocelli, II, *Medicine: An Illustrated History* (New York: Abradale Press, 1987), 105.
3   Roy Porter, ed., *Medicine: A History of Healing* (New York: Marlowe and Company, 1997), 174.
4   G.P. Srivastava, *History of Indian Pharmacy*, 180.
5   Roy Porter, *The Greatest Benefit to Mankind: A Medical History of Humanity* (New York: Norton, 1997), 137. M. Patricia Donahue, *Nursing*, 48.
6   Ibid. (Donahue), 48.
7   G.P. Srivastava, *History of Indian Pharmacy*, 185.
8   Ibid., 185.
9   G.P. Srivastava, *History of Indian Pharmacy*, xi, foreword.
10 Albert S. Lyons and R. Joseph Petrocelli, II, *Medicine*, 113.
11 Roy Porter, *The Greatest Benefit to Mankind*, 141.
12 G.P. Srivastava, *History of Indian Medicine*, 128.
13 Ibid., 141.
14 Ibid., 141–145.
15 Ibid., 240.
16 Ibid., 126.
17 Roy Porter, *The Greatest Benefit to Mankind*, 143.
18 Ibid., 144–145.
19 Deepak Chopra, *Perfect Health* (New York; Harmony Books, 1990), 21–95.
20 Rick Smollan, Phillip Moffitt, and Matthew Naythons, *The Power to Heal* (New York: Prentice Hall Press, 1990), 19.
21 Albert S. Lyons and R. Joseph Petrucelli, II, *Medicine*, 114.
22 John C. Hartnett, "The Care and Use of Medicinal Leeches," *Pharmacy in History*, Vol. 14 (1972), No. 4, 127–138.
23 G.P. Srivastava, *History of Indian Medicine*, 191–192.
24 Lois N. Magner, *A History of Medicine* (New York: Marcel Dekker, Inc., 1992), 48.
25 Roy Porter, *The Greatest Benefit to Mankind*, 152.
26 Lois N. Magner, *A History of Medicine*, 53.
27 Manfred Pokert with Christian Ullmann, *Chinese Medicine*, trans. Mark Howser (New York: William Morrow & Company, 1988), 74.
28 Lois N. Magner, *A History of Medicine*, 48.
29 Erwin Ackerknecht, *A Short History of Medicine* (Baltimore, MD: The Johns Hopkins University Press, 1982), 44.

30 Joel L. Swerdlow, *Nature's Medicine: Plants That Heal* (Washington, DC: National Geographic Society, 2000), 37. Paul Unschuld, *Medicine in China: A History of Pharmaceutics* (Berkeley: University of California Press, 1986), 11.

31 Lois N. Magner, *A History of Medicine*, 52. Manfred Pokert, *Chinese Medicine*, 83.

32 Ibid., Magner, 52.

33 Cedric B. Baker, "The Historical Evolution of Natural Pharmacy: Medicinal Foods and Phytomedicines in Asia & Europe" (Powerpoint Presentation Lecture at Cornell University, Ithaca, NY, September 9, 2010).

34 Manfred Pokert, *Chinese Medicine*, 254–255. Albert S. Lyons and R. Joseph Petrucelli, II, *Medicine* (New York: Abradale, 1987), 124.

35 Cedric B. Baker, "The Historical Evolution of Natural Pharmacy."

36 Manfred Pokert, *Chinese Medicine*, 257–258.

37 Ibid., 258.

38 Lois N. Magner, *A History of Medicine*, 57.

39 Manfred Pokert, *Chinese Medicine*, 31.

40 Ibid., 247.

41 Ibid., 249.

42 Ibid., 252.

43 The author would like to acknowledge the contributions of Dr. Cedric Baker, Pharm.D. of Mercer University for his careful reading of this chapter and for his sagacious comments and contributions to making it much better than it otherwise would have been.

4 Pharmacy in Ancient Greece and Rome

How did Greco-Roman philosophy come to dominate Western pharmacy and medicine for the next 1,400 years?

Ancient Greece

Although the height of their influence lasted from 500 BCE to about 500 CE, the ancient Greeks set the standard for Western pharmacy and medicine until germ theory was established scientifically in the late nineteenth century. How did this come to be? The short answer is the birth of free thought began relatively unencumbered in ancient Greece. The geography of Greece had a profound impact upon its cultural development. The Greek peninsula was replete with rocky hills composed of chalky soil that compelled the Greeks to look outward from their land to trade with other civilizations for the food they could not grow themselves. Most notably, there was no single river system that connected the hundreds of city-states and islands, which allowed them to grow independently from one another. Without a strong central government or a centralized priestly caste, the ancient Greeks were free to generate many forms of thought and experiment with them. This freedom of thought allowed such talented outliers such as Hippocrates of Cos (*c.*460–377 BCE) and Pedanios Dioscorides (*c.*40–90 CE) to transform the practice of medicine and pharmacy, respectively.

Mythical Pharmacy and Medicine

The earliest ancient Greeks who were the descendants of the Minoans and Dorian warriors defined health and illness in mythological terms. Apollo, sometimes called Alexikakos (the averter of ills), the god of disease and healing, could inflict illness on mortals by shooting poisoned arrows into them. Nonetheless, it was Chiron the Centaur who first mastered the art of pharmacy according to disparate sources including Homer and Pliny.[1] In turn, Chiron taught the art of pharmacy to Asclepius, Achilles, Jason, Odysseus, and others. Apollo's son Asclepius became the most significant healer who was depicted carrying a snake entwined staff, called a caduceus, which survives as a symbol of both the pharmacy and medical professions. Asclepius and his wife Epione had many children; the most notable were Hygeia, who symbolized health, and Panacea who became a symbol for medicine.[2] Hygeia often has been depicted with a mixing bowl named in her honor and it has become a symbol for pharmacy in the Netherlands. Today, in the United States, the Bowl of Hygeia Award is a prestigious honor awarded to an outstanding pharmacist from each state on an annual basis.

The Temples of Asclepius

Similar to his counterpart in Egypt, Imhotep, who replaced Thoth as the god of medicine, Asclepius eventually eclipsed Apollo as the god of medicine. By 500 CE, there were nearly 200 Asclepian temples in the ancient Greco-Roman world, patterned after the Temple at Epidaurus. The Temple at Epidaurus resembled a modern spa and had a theater, a stadium, a hotel, gymnasia, and baths surrounded by a peaceful forest. Over time, the temples became centers of worship, civic activity, and recreation for the sick and healthy alike. The temples had staffs composed of priests, priestesses, musicians, singers, and others as needed to operate the facilities. Sacred dogs and snakes roamed the grounds. Dogs were encouraged to lick the open wounds of patients as a form of therapy. Today, dog saliva has been found to have some antibacterial properties. Snake venom might have been used to induce hallucinations during the patients' incubation sleep.[3] Patients presented their symptoms to the priests who would prescribe various exercises, diets, rest, or drugs. Patients were directed to sleep in a part of the temple known as a *kline*, which gives us the modern term "clinic."[4] Patients slept in the temple at night, a process known as incubation, and were encouraged to tell the priests about their dreams and the priests would, in turn, interpret the dreams for the patient and offer treatment. To the degree that a patient could remain active, they were encouraged to stay engaged in activities that would take their mind off their illness—perhaps this was the beginning of occupational and musical therapy. One form of therapy called for patients to fashion plaster models of their diseased or injured body part and present it to the altar of Asclepius as a token of gratitude for healing them. These plaster models were known as votives. The votives were put on display and new patients could see tangible evidence of healing as they entered the temple. The Romans would adopt this model of healing and the temples continued to operate until the Middle Ages.[5]

Gradually, the pre-Socratic philosophers used the power of reason to explore their natural world and developed more rational explanations that defined the ancient Greek theory of health and illness. Empedocles of Agrigentum (*c*.500–430 BCE) developed an ancient version of the periodic

Figure 4.1 Asclepius was a semi-divine legendary ancient Greek healer who was known
for his healing power. By 500 CE, there were over 200 Temples of Asclepius
built throughout the Western ancient world.

Source: Vatican Museum. Photo: Alinari/Art Resource, NY.

table that classified all matter into four categories: earth, air, fire, and water.[6]
Around 400 BCE, Hippocrates of Cos speculated that the human body was
composed of four humors (fluids) that corresponded to these four categories
of matter and included black bile, blood, yellow bile, and phlegm. These four

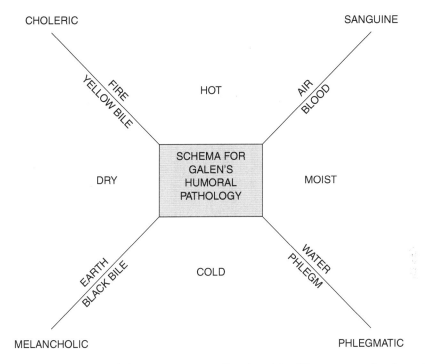

Figure 4.2 Diagram of the Four Humors—Developed by Hippocrates and enhanced by Galen, the four bodily humors in balance represent a healthy body. Any imbalance in this perfectly balanced system signified illness. Humoral theory dominated Western medicine for nearly 1500 years until the discovery of germ theory in the 1870's.

humors also exhibited qualities associated with the four categories of matter. For example, black bile, like earth, was cold and dry. Blood, like air, was moist and hot. Yellow bile, like fire, was dry and hot. Phlegm, like water, was moist and cold. This squared the philosophical circle that the human body was a microcosm of nature. Health meant that these humors were in perfect balance. By contrast, too much or too little of any of these humors signified illness. This logical theory of health and illness was consistent with the Greek concept of *sophrosyne*, which meant balance in all things. Most importantly, Hippocrates maintained that health and illness could be explained in natural, and not supernatural terms.[7]

Hippocrates "The Father of Medicine"

Sometime between 480 and 380 BCE, Hippocrates (*c.*460–377 BCE) and his associates wrote the *Corpus Hippocraticum*, a medical encyclopedia of sorts that mentions over 300 different herbal remedies. In accordance with humoral theory, Hippocrates and his followers relied greatly on diet, exercise, proper rest, and behavioral therapies to treat illness. The treatments of last resort were drug therapy or minor surgical procedures that might include bleeding, enemas, administering diuretics, sweating, and vomiting. In his treatise on *Airs*, Hippocrates noted that "contraries are the cure of contraries." For example, if a patient was suffering from a cold, which was a phlegmatic condition exhibiting cold and moist qualities, the physician might administer a mixture of cumin and hyssop, which had hot and dry qualities, to restore the proper humoral balance to the patient. This treatment by opposites became known as allopathic medicine, which has become the dominant form of medicine practiced in the world today. Thus, ancient Greek pharmacy, and consequently Western pharmacy, were influenced profoundly by humoral theory.[8]

Ancient Greek Medical Sects

After Hippocrates' death, the longstanding philosophical tensions among Greek physicians became apparent as each sect sought to impose their ideas at the expense of the others. Prior to Hippocrates, Greek physicians subscribed to one of two schools of medical thought. The Cnidus school stressed the *diagnosis* of disease, i.e. observing a patient's symptoms and identifying the disease, and then treating the symptoms with remedies such as drugs or minor medical procedures. The Cos school, which Hippocrates would advance, stressed *prognosis*, or forecasting the course a disease would take by observing the patient's condition and relying on the healing powers of nature. This accounts for the Hippocratic Oath's most famous advice for healers, "to do no harm." In other words, the physicians of the Cnidus school would treat a disease more aggressively; whereas, the physicians of the Cos school would be more apt to treat the patient with the least invasive natural treatments such as diet, sleep, and exercise. The Cos school's emphasis on prognosis harkened back to the Mesopotamian reliance on astrology, but instead of relying on divine omens for the guidance of treatment, Greek physicians relied on the patient's physical symptoms and condition. The Cnidians' emphasis on diagnosis resembled the Egyptian, Indian, and Persian traditions of treating the symptoms of the disease rather than the patient in a rather mechanical therapeutic manner.[9]

After Hippocrates' death, medical sectarianism increased among Greek physicians. While there were a multitude of sects, the two most historically significant were the Rationalists and the Empirics. The Rationalists believed that sound medicine had to rely on the study of human anatomy and physiology, and for a brief time during the Alexandrian era, made great advancements by dissecting humans. On the contrary, the Empirics, founded by Heraclides, rejected anatomy and physiology as the basis of medicine and argued that curing disease was all that really mattered. The Empirics believed in three principles for practicing successful medicine that consisted of observation, history, and analogy. Ultimately, the Empirics won and this resulted in medicine being reduced to a mechanical system of restoring bodily humors, usually through the administration of poly-pharmaceuticals. Treating classified sets of symptoms in this way made the humoral theory of illness the central tenet of Western medical practice until it was displaced by germ theory in the late nineteenth century. For pharmacy, the victory of the Empirics had profound ramifications on shaping the future course of Western pharmacy with its emphasis on searching for poly-pharmaceutical panaceas.[10]

The Founding of Botany

Diocles of Carystus, a late fourth century BCE Athenian *rhizotomoi*, or root collector, expanded on Hippocrates' work by compiling a herbal guide that emphasized the medicinal value of plants in treating patients.[11] Theophrastus of Eresus (*c.*370–287 BCE), one of Aristotle's greatest students, built upon Diocles' work and founded the field of biology known as botany. In his works *On the*

Figure 4.3 Theophrastus (*c.*300 BCE), a student of Aristotle, became the founder of the field of botany in ancient Greece.

Source: Painting by Robert Thom, reproduced courtesy of the American Pharmacists Association Foundation.

Theophrastus wrote a treatise about perfumes during Alexander the Great's time. In fact, Alexander the Great had exotic plants sent back to Theophrastus in Athens for study. In his treatise, Theophrastus observed that light rose-based perfumes such as kyros that had delicate fragrances were best suited for men; while, by contrast heavier, more persistent scents such as Megalion, derived from myrrh oil, sweet marjoram, and spikenard were best suited for women. Theophrastus discussed the use of rose petal-based pot-pourri to scent clothes and powdered fragrances that could be used to sprinkle in bedding.[12]

Natural History of Plants and *On the Origins of Plants*, Theophrastus identified and classified over 500 species of plants and explained their medicinal properties. He succeeded Aristotle at the Lyceum and taught there for nearly 35 years. He has often been called "the father of botany" and perhaps deserves the title "the father of pharmacognosy" as well. Remarkably, Theophrastus pioneered the concept of drug tolerance, observing that the power of a drug taken over a long period diminishes in people who become accustomed to taking them.[13]

The Founding of Toxicology

Nicander of Colophon (*c.*125 BCE) was a talented physician and poet who lived during the reign of King Attalus III of Pergamon in Asia Minor. He wrote two volumes of verse about antidotes to poisons entitled *Theriaca* and *Alexipharmaca*. The first poem deals with the treatment of venomous serpents and animals, while the second poem deals with antidotes to poisons that have been ingested. The word "theriac" derives from the Greek word *therion* meaning wild venomous beast.[14] Similarly, the word "antidote" derives from the Greek word *antidotos* meaning "giving against." The idea was to antagonize one poison with another to affect a cure.[15] While not well known himself, Nicander's work influenced the advancement of toxicology by King Mithridates VI of Pontus, Galen, Pliny, and others.[16]

King Mithridates VI of Pontus: The "Father of Toxicology"

One of the more unlikely people to advance pharmacy was King Mithridates VI who ruled over Pontus from about 120–63 BCE. By all accounts, Mithridates had a classical Greek education and became interested in poisons and their antidotes due to the intrigues that plagued many royal courts during this era. Mithridates spent his life opposing the expansion of the Roman Empire into his native Asia Minor and for that reason faced many enemies. He employed many *rhizotomists* (root collectors) and *pharmacopolists* (drug sellers) at his court. The King personally experimented with poisons and their antidotes by testing them on ducks, live prisoners, courtiers, and then on himself. His most famous antidote against all poisons was a poly-pharmaceutical called *Mithridatium* that contained over 70 ingredients including: gentian, valerian root, cinnamon, opium, blood from ducks that had been raised on toxic plants, viper flesh (representing Asclepius), and skink (small salamander). The idea behind poly-pharmaceuticals was that if one ingredient did not help the patient another ingredient would. Over the next millennium *Mithridatium* became regarded as a panacea for all sorts of ailments and with some minor modification took on several names including: treacle, galene, and theriac. For his achievement, King Mithridates has become known as "the father of toxicology." Ironically, before his palace was overrun by the Roman General Pompey's forces, Mithridates tried to poison himself, but none of the poisons worked. So, he ordered one of his soldiers to kill him, which the soldier did. General Pompey captured the recipe for *Mithridatium* and brought it back to Rome where it was studied by physicians including Claudius Galen, and the panacea lasted for nearly a millennium. *Mithridatium* was reformulated by Emperor Nero's physician Andromachus. In fact, *Mithridatium* was produced until at least 1752 CE when it was expunged from the pharmacopeia of the

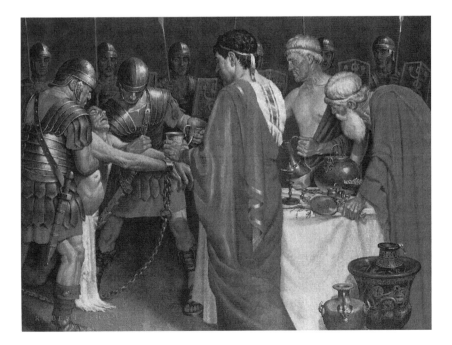

Figure 4.4 King Mithridates VI of Pontus (*c.*100 BCE) concerned about being assassinated by poisoning, became the founder of the field of toxicology.

Source: Painting by Robert Thom, reproduced courtesy of the American Pharmacists Association Foundation.

Royal College of Physicians of Edinburgh. In modified form, it was mentioned in the French Codex of 1866, and it continued to be used in Turkey into the twentieth century.[17]

Terra Sigillata: The First Trademark Drug

In addition to *Mithridatium* there were two other great remedies used by ancient Greco-Roman physicians that have survived for nearly 2,000 years: *Terra Sigillata* (Sacred Sealed Earth) and *Hiera Picra* (Holy Bitter). *Terra Sigillata* was first discovered on the Greek island of Lemnos and was a white clay substance that was dug out of the side of a hill, according to legend, one time per year to pay homage to the goddess Diana. Once harvested, the clay was washed, dried, rolled to proper thickness, cut into lozenges, and then impressed with an official seal by a priestess. Thus, *Terra Sigillata* became the first trademarked drug that was sold commercially, ensuring that patients were

Figure 4.5 Terra Sigillata (c.500 BCE) or "sealed earth" from the ancient Greek island of Lemnos became the first trademarked drug. Rolled into a flat clay pastille, *Terra Sigillata* was imprinted with an insignia as symbol of authenticity.

Source: Painting by Robert Thom, reproduced courtesy of the American Pharmacists Association Foundation.

buying an authentic product from an established producer. While the formula has changed over the millennia, the basic ingredients of the drug include greasy clay, silica, aluminum, chalk, magnesia, iron oxide, and according to legend, goat's blood. In ancient times it was used for insect and animal bites, watery eyes, dysentery, fevers, and poisons. *Terra Sigillata* and several of its competitors lasted well into the nineteenth century and was produced well into the 1890s.[18]

Hiera Picra

Hiera Picra or Holy Bitter dates back to the time of the Asclepian temples. Its principal ingredients were aloes and canella, but over time other ingredients were added such as colocynth. For example, Themison of Laodicea's formula from about 43 BCE included aloes, mastic, saffron, Indian nard, carpobalsam, and asarum. Another very popular formula came from Antonius Pachius who kept it secret until his death, whereupon Scribonius Largus studied his papers and reported the ingredients as follows: colocynth, agaric, germander, white horehound, lavender, opoponax, sagapenum, parsley seeds, round birthwort root, white pepper, spikenard, cinnamon, myrrh, and saffron. Since Holy Bitter had an overpowering taste as a powder, it was mixed with honey, treacle, or some other vehicle to make it palatable. Similar to *Mithridatium* and *Terra Sigillata*, *Hiera Picra* was modified by various apothecaries and physicians over the centuries. *Hiera Picra* was included in the United States Pharmacopeia until 1870. Use of the ancient drug continued into the twentieth century in Belgium and rural England.[19]

Ancient Greek Dosage Forms

The ancient Greeks liked to administer their medicines in confections or electuaries. Confections were made by mixing powders with fruits or honey in a mortar until it formed a soft mass and then was stored in clay pots or jars. When needed, the confections were removed from the jars with a spatula and formed into a round ball ready for swallowing. These balls were called *katapotia* meaning "things to be swallowed." Later during Roman times, Pliny mentions the word *pilula* referring to a small amount of medication. Some scholars suggest that this might be the origin of our word for pill.

An electuary was a thinner and less congealed form of a confection that was taken by licking the substance from a spoon or spatula. Sometimes an electuary was smeared onto a licorice root or a twig for consumption. The Greek word *ekleikton* literally means "something to be licked" and gives us the word *electuary*. Over time, confection, electuary, and theriac were often used interchangeably, only adding to the apothecary's confusion in preparing poly-pharmaceuticals.[20]

The Roman Era of Pharmacy and Medicine

The Transition From Greek to Roman Medicine

As a civilization, the ancient Greeks were innovative and speculative and provided Western civilization with a system of allopathic medicine based on humoral theory. The very system of fiercely independent city-states that promoted free thinking and the tremendous cultural achievements that followed also caused internal conflicts and rivalries among the city-states that led them to their own destruction—first with the Peloponnesian War, the conquests of Alexander the Great, and then their complete conquest by the Romans. While the Greeks gave us the Hippocratic Oath and humoral theory, the Romans gave us public health insurance (a system in which citizens paid special taxes to support local physicians who would treat them), systematic pharmacopeias, and the best military hospitals the world would see until the twentieth century. In contrast to the Greeks, the Romans as a civilization were highly practical and valued the classification, organization, and the application of ideas. When the Romans conquered the Greeks they borrowed many Greek ideas and institutions and improved upon them. It should come as no surprise then that many of the key innovators in what would become the Greco-Roman era in the history of pharmacy and medicine were of Greek ancestry. Thus, the Greco-Roman era was born.

Pedanios Dioscorides and the First Western Materia Medica

The creator of the first authoritative *materia medica* or pharmacopeia, Pedanios Dioscorides (*c.*40–90 CE) served as a surgeon to Roman legions, which afforded him the opportunity to travel and to study plants. Dioscorides traveled widely throughout the ancient world with the Roman legions from Gaul to his native Asia Minor collecting, studying, testing, categorizing, and describing the properties of the specimens he found that had medicinal value.[21] His *Herbarium De Materia Medica* described about 600 plants, 35 animal products, and 90 minerals that could be used to heal patients. He discovered about 100 more plants than Theophrastus and knew about 450 more plants than Hippocrates did. About 90 of the substances that Dioscorides studied are still used in modern pharmacy, including: aloes, asafotida, cinnamon, copper sulfate, gum ammoniacum, horehound, lanolin, male-fern, myrrh, mercury, opium, ox-gall, and squill to name just a few.[22] Dioscorides basically set the format for future pharmacopeias. Each item was listed with its synonyms, habitat, botanical description, medicinal properties, medical usage, dosage instructions, harvesting procedures, preparation, storage, adulteration, and tests for adulteration. There were illustrations of the herb next to each description. The greatest proof of the value of Dioscorides' work is that it was widely circulated in the West in Greek, Latin, and Arabic and remained in continuous use until 1600 CE.

Aulus Cornelius Celsus: Roman Encyclopedist

True to the Roman devotion to organization, Aulus Cornelius Celsus (*c*.25–50 CE) was probably not a physician or an apothecary, but a talented scholar and translator who wrote perhaps the best medical encyclopedia of antiquity, entitled *De re Medicina*. The original encyclopedia dealt with many subjects other than medicine, but *De re Medicina* was the only part to survive over the next millennia. *De re Medicina* was divided into eight books dealing with topics including: the history of medicine, general pathology, disease states, anatomy, pharmacy, surgery, and orthopedics. A prescient writer, Celsus coined the term *insania* giving us the modern term "insanity." Similarly, he mentions heart disease, which he called *cardiacus*. The fifth book on pharmacy consists of a list of drugs, with chapters on weights and measures, compounding prescriptions, and the therapeutic use of drugs.[23] In terms of pharmacy, Celsus wrote about drug preparations involving opiates and even suggested a prophylactic for tooth decay called *sory*. *Sory* consisted of poppy seed, pepper, and copper sulfate compounded into a paste with galbanum.[24]

In the fifth book of *De re Medicina*, Celsus describes cataplasms (poultices) that were made by boiling flour in water rendering a thick paste which was mixed with gums or wax. Some cataplasms were applied to wounds, while others called *malagma* were applied to unbroken skin to provide warmth for internal disorders and over abscesses and swollen joints. Troches and pastilles (lozenges) were made by taking dry ingredients and mixing them with a liquid such as wine or vinegar. Once the mixture reached the desired consistency it was molded (usually flat and round) and left to dry. Celsus also described catapotia which was a painkiller composed of mandrake, henbane seed, wild rue, and opium. It was powdered and then shaped into a small mass. Conserves were medicines made by combining their powder with boiled honey. Collyria were medicines made to be applied to the eyes. Celsus describes the making of suppositories and the Romans were masters in the making of ointments (cerates) as evidenced by Galen's formulation of cold cream.[25] By any measure, Celsus captured the essence of Roman pharmacy.

In his book, Celsus also described the four cardinal symptoms of inflammation: *calor* (warmth), *dolor* (pain), *rubor* (redness), and *tumor* (swelling). In his translation of Hippocrates' Greek word *carcinos*, Celsus translated the word into the Latin word *cancer* meaning crab. Having survived the collapse of antiquity and the Middle Ages, Celsus' *De re Medicina* was one of the first books printed after the Bible during the printing revolution launched by Johan Gutenberg in the fifteenth century. It was printed in 1478, the first medical book to be so honored.[26]

Pliny the Elder

Another, better known Roman encyclopedist was Pliny the Elder (*c*.23–79 CE) whose work *Historia Naturalis* systematically organized Greco-Roman knowledge into 37 books, many dealing with botany and the medicinal use of plants. Pliny argued that plants existed to serve human needs. Pliny also introduced the concept of the "doctrine of signatures" in the West that maintained that plants exhibited characteristics in terms of color or form that were similar to the conditions they might be used to treat. For example, the lungwort plant is a plant that sports lush leaves that resemble the human lungs, and consequently was used to treat respiratory ailments. Much of Pliny's work perpetuated folklore and myths about the magical healing qualities of some plants that persisted into the twentieth century. Pliny's *Natural History* continued to exert considerable influence on Western pharmacy until the seventeenth century. Being a curious naturalist, Pliny died investigating the eruption of Mount Vesuvius when he was overcome by the toxic fumes from the volcano.[27]

Scribonius Largus

Scribonius Largus (*c*.1–50 CE) was a prominent Roman physician who wrote a practical handbook of 271 drug recipes known as *Compositiones* (or *Prescriptions* in modern English). A product of his time, Largus believed in using poly-pharmaceuticals in treating patients. He described 242 plants, 36 minerals, and 27 animal products that he used in preparing his prescriptions. His work also provides a further glimpse into the Roman practice of pharmacy.[28]

Claudius Galen

Although Hippocrates was known as the "father of medicine," in terms of sheer influence Claudius Galen of Pergamum (130–200 CE) surpassed him. Attending physician to the Roman gladiators and later to three Roman emperors, Galen shaped the course of Western pharmacy and medicine for over a millennium, until the discovery of germ theory. Galen's influence came due to his prolific writing on medical and pharmacy topics. In his book *On the Art of Healing* he described 473 drugs that he used to treat his patients. Consistent with the practices of his era, Galen prepared his own medications and criticized physicians who allowed others to do so. Galen prepared and administered his drugs in an area known as an *iatreon* and stored them in another area called an *apotheca*. Galen prepared his own version of *Mithridatium* that consisted of viper flesh, herbs, wine, minerals, and honey he called *galene*.[29] For toothaches, he prepared a prescription that included black pepper, saffron, opium, carrot seeds, aniseed, and parsley seed compounded into a paste that was placed in the tooth cavity.

He used henbane seeds, mandragora, and poppy juice as painkillers and used colocynth, elaterium, hellebore, and scamony as purgatives in his practice. Perhaps his most lasting contribution was the invention of cold cream that was known as *unguentum* or *ceratumrefrigerans*, a basic ingredient in many modern cosmetics. These preparations of vegetable ingredients that produce no actual chemical changes are still referred to as *galenicals*.

Marcellus Empiricus

A key Roman medical writer from Gaul, Marcellus Empiricus, was a transitional writer who wrote a *materia medica* which, while grounded in Greco-Roman medicine and pharmacy, foreshadowed an approach to both fields that would become the standard in medieval writing. Marcellus Empiricus, sometimes known as Marcellus Burdigalensis, lived around 400 CE and served as a high official under the Roman Emperor Theodosius I. His treatise entitled *De medicamentis* was a compendium of letters from Greco-Roman practitioners followed by a series of drug treatments arranged anatomically from head to toe. His work drew on the works of Celsus and Pliny and also included Celtic folk treatments as well as magical explanations. While not as influential as Dioscorides' or Celsus' work, *De medicamentis* signaled a transition from the emphasis of Mediterranean influences to the newer world of Western and Northern European pharmacy. Two strands of pharmacy began to emerge at this time: the mainstream traditional pharmacy promoted by Galen and a more folk medicine, herbal tradition begun by Scribonius Largus. Marcellus Empiricus' work would follow Largus' school of herbal remedies.[30]

Pharmacy Practice in Ancient Rome

Many physicians in Roman times followed Galen's advice and prepared their own medications. Nonetheless the Latin language indicates that there was a division of labor and specialization when it came to the preparation of drugs. *Medicina* were the drugs themselves that were prepared by people called *medicamentarii, pharmacopoei, pharmacotritae,*or *confectionarii.* The term *medicamentarius* could also describe a person who administers poison. There were also makers and sellers of ointments, known as *ungentarii* and *seplasarii,* respectively. In Roman market places it was fairly common to see preparers and vendors of spices, called *aromatarii* as well as sellers of dyes, known as *pigmentarii.*[31]

Being a highly commercial civilization, the Romans also made distinctions in describing drug sellers. *Seplasia* was the Roman name for a pharmacy. Drug vendors who kept their own shops or street stalls were referred to as *selluarii. Pharmacopolae circumforaneae* were traveling drug salesmen. Pliny, in his writing, chastised physicians for purchasing drugs from *seplasarii* without checking into the ingredients and how the drugs were prepared. A similar situation emerged in the United States during the nineteenth century when the American Medical Association warned physicians to be wary in prescribing potentially dangerous patent medicines to unsuspecting patients.[32]

The Origin of Rx

The Romans had a profound impact on the practice of pharmacy as demonstrated by the basic terms pharmacists continue to use today. For example, the symbol *Rx* meaning prescription derives from the Latin word *recipere* meaning to prepare. According to legend, the *Rx* symbol came into use during the reign of the Roman Emperor Nero near the start of the Christian Era when the Romans persecuted the Christians. In order to show loyalty to Nero, the Roman Empire, and the Roman religion, physicians started using the abbreviation for Jupiter 4, which was *Rx*, on the top of all of their prescriptions and the practice has continued to the present day.[33]

Another symbol of pharmacy that the Romans named for us was the mortar and pestle. Although mortars and pestles date back at least as far as the ancient Egyptians, the Romans gave us their name. The Latin term *mortarium* meant a "receptacle for grinding" which comes down into English as *mortar*. Similarly, the Latin word *pestillum* evolved into the English word *pestle*. The earliest mortars and pestles were made of wood, fired clay, or stone and then metal as metallurgy evolved. The Romans also used a type of grinding device known as a quern. Dioscorides and Galen also introduced the apothecaries' system of weights and measures in preparing drugs.[34]

Chapter Summary

Pharmacy and medicine reached new heights of achievement under the Greeks and especially under the Romans. The Greeks provided the West with a humoral theory of health and illness that would persist until germ theory was discovered in the late nineteenth century. Allopathic medicine still dominates medical treatment in the world today. Upon graduating from pharmacy or medical schools, pharmacists and physicians take oaths based on the ethical precepts introduced by Hippocrates that they will "do no harm" to patients and to protect and act in the best interest of their patients.

Key Terms

Apollo	Nicander of Colophon	Scribonius Largus
Chiron	Mithridates VI	Claudius Galen
Asclepius	*Mithridatium*	galenicals
Hygeia	*Terra Sigillata*	Rx
Panacea	*Hiera Picra*	electuary
Empedocles	Pedanius	troche
Hippocrates	Dioscorides	pastille
allopathic medicine	Aulus Cornelius Celsus	
Theophrastus	Pliny the Elder	

Chapter in Review

1   Explain the origins of humoral theory and how it affected the development of pharmacy.
2   Explain how the Romans built upon Greek medical and pharmacy achievements.
3   Explain what a patient experienced at a Temple of Asclepius.

Notes

1   Charles Lawall, *The Curious Lore of Drugs and Medicines: Four Thousand Years of Pharmacy* (Garden City, NY: Garden City Publishing, Inc., 1927), 32. Hermann Peters, *The Pictorial History of Ancient Pharmacy with Sketches of Early Medical Practice*, trans. William Netter (Chicago, IL: Engelhard & Company, 1889), 7. (Reprint, Lexington, KT: Forgotten Books, 2011) www.forgottenbooks.com.
2   Jeanne Achterberg, *Woman as Healer* (Boston, MA: Shambala Publications, Inc., 1990), 30–31.
3   Ibid., 30.
4   Ann Rooney, *The Story of Medicine* (London, Arcturus Publishing Limited, 2009), 184.
5   Fielding H. Garrison, *An Introduction to the History of Medicine* (Philadelphia: W.B. Saunders Company, 1913), 83–84.
6   Roberto Margotta, *The Story of Medicine* (New York: Golden Press, 1968), 25.
7   C.G. Cumiston, *The History of Medicine* (New York: Dorset Press, 1987), 100–101.
8   Erwin Ackerknecht, *A Short History of Medicine* (Baltimore, MD: The Johns Hopkins University Press, 1982), 53.
9   James Grier, *A History of Pharmacy* (London: The Pharmaceutical Press, 1937), 24.
10 Ibid., 25.
11 David Cowen and William H. Helfand, *Pharmacy: An Illustrated History* (New York: Harry Abrams, 1990), 11.
12 Sally Pointer, *The Artifice of Beauty* (Gloucester: Sutton Publishing, Ltd., 2005), 42–43.
13 D.C.A. Hillman, *The Chemical Muse* (New York: St. Martin's Press, 2008), 63.
14 James Grier, *A History of Pharmacy*, 50.
15 Ibid., 254.
16 John Scarborough, "Nicander's Toxicology," *Pharmacy in History* 21 (1979), 3. Joseph B. Sprowls, Jr. and Harold M. Beal, eds., *American Pharmacy: An Introduction to Pharmaceutical Technics and Dosage Forms* (6th ed., Philadelphia: J.B. Lippincott Company, 1966), 7.
17 Adrienne Mayor, *The Poison King: The Life and Legend of Mithradates, Rome's Deadliest Enemy* (Princeton, NJ: Princeton University Press, 2009), 237–243. C.J.S. Thompson, *The Mystery and Art of the Apothecary* (London: John Lane the Bodley Head, Ltd., 1929), 66.
18 Ibid. (Thompson), 55. Charles Lawall, *The Curious Lore of Drugs and Medicines*, 69. A.C. Wootton, *Chronicles of Pharmacy* (Reprint: Boston, MA: USV Pharmaceutical Corp., 1972) Vol. 2, 53–54.
19 C.J.S. Thompson, *The Mystery and Art of the Apothecary*, 42.
20 Joseph B. Sprowls, Jr. and Harold M. Beal, eds., *American Pharmacy*, 8.
21 John M. Riddle, *Dioscorides on Pharmacy and Medicine* (Austin: University of Texas, 1985), 1–6. John Scarborough, "Roman Pharmacy and the Eastern Drug Trade," *Pharmacy in History*, Vol. 24 (1982), No. 4, 135–140.

22 David L. Cowen and William H. Helfand, *Pharmacy*, 28. Fielding H. Garrison, *An Introduction to the History of Medicine* (Philadelphia: W.B. Saunders Company, 1913), 109.

23 Ibid., 108.

24 Charles Lawall, *The Curious Lore of Drugs and Medicines*, 56.

25 C.J.S. Thompson, *The Mystery and Art of the Apothecary*, 28–29. Joseph B. Sprowl, Jr. and Harold M. Beal, eds., *American Pharmacy*, 8–9.

26 Fielding H. Garrison, *An Introduction to the History of Medicine*, 107.

27 Ibid., 112.

28 D.C.A. Hillman, *The Chemical Muse*, 41.

29 David L. Cowen and William H. Helfand, *Pharmacy*, 34.

30 C.J.S. Thompson, *The Mystery and Art of the Apothecary*, 68.

31 Ibid., 36.

32 Charles Lawall, *The Curious Lore of Drugs and Medicines*, 61.

33 Otto A. Wall, *The Prescription: Therapeutically, Pharmaceutically, Grammatically, and Historically Considered* (St. Louis, MO: August Gast Bank-Note and Litho. Company, 1898), 200–206.

34 Charles Lawall, *The Curious Lore of Drugs and Medicines*, 61.

5   Medieval Pharmacy in the West
Pharmacy in an Age of Faith
(Dogma and Authority)

How did pharmacy survive and advance during the medieval era?

Just as Greco-Roman pharmacy and medicine had lasted from 500 BCE to 500 CE, medieval pharmacy and medicine would last for nearly a millennium from 500 CE to 1500 CE. In the wake of the collapse of the Roman Empire in 476 CE, the whole course of Western civilization seemed to reverse itself. In the political power vacuum that occurred and the economic displacement that followed Rome's collapse, people fled the cities seeking protection in the countryside from feudal warlords. Conditions were so desperate that people were willing to sell themselves and their families into lives of perpetual serfdom in exchange for a modicum of protection from a warlord. Amid this unprecedented state of chaos the medieval Church filled the power vacuum and fostered a new world view that explained human health, illness, and suffering in Judeo-Christian terms. The medieval era was, above all else, an age of faith and while faith-based treatments existed in the ancient world, faith would play a central role in the development of medieval pharmacy and medicine.

The medieval Church attempted to become a totally dominant institution and pharmacy and medicine thus became extensions of Church doctrine. Dating back to the ancient Hebrew tradition, illness was seen as a physical manifestation of sin leading to suffering, and therefore the treatment was prayer and repentance in search of divine grace which held out hope for healing. While illness was a mark of sin, the altruistic act of caring for the sick became a devotional act of faith for Jews, Christians, and Muslims.[1] The best known example of faith-healing was that of Jesus and the apostles who supported themselves by traveling around Palestine healing the sick. St. Luke was of course a physician and became an inspiring example to many during the desperate times in the Middle Ages, becoming a patron saint to physicians.

Saintly Pharmacy

Another powerful example came in the form of the twin Arab brothers from Asia Minor known as Damian and Cosmas. Damian was an apothecary and Cosmas was a physician. They traveled together caring for the sick and eager to spread their Christian faith, accepting no fees for the treatment they offered. In the fourth century CE the Roman Emperor, Diocletian, launched a campaign to persecute the Christians and the twin brothers were arrested, tortured, and executed in February 303 CE for refusing to renounce their faith. They were canonized and became the patron saints of pharmacy and medicine, respectively.[2]

From these examples hospitals were built as acts of Christian charity. The Roman Emperor Constantine's mother St. Helena opened a hospital in Constantinople in 330 CE, followed by St. Basil, who opened a hospital in Caesarea, the capital of Cappadocia. St. Epharem founded a hospital for plague victims at Edessa in Asia Minor, and Fabiola, a wealthy Roman woman who studied under St. Jerome, founded the first Christian public hospital in Europe around 394 CE.[3]

Faith-Healing and the Placebo Effect

In the scientific community today where knowledge must be empirically validated by use of the scientific method, faith-healing and much of what was done to heal people in the ancient world might be easily dismissed, but one must place one's self in the historical circumstances to understand why these people believed in the things they did. For example, one might argue that much of the healing that occurred might be attributed to the placebo effect.[4] Studies about the power of the placebo effect in recent years have yielded some startling results. For example, a national survey conducted among practicing physicians in the US in 2008 revealed that over half of the physicians surveyed prescribed "placebo" drugs to over half of their patients and, more importantly, most physicians surveyed did not see anything wrong with that practice.[5] The placebo drugs involved are often vitamins or mild antibiotics. Pharmacists are also complicit in this practice by dispensing these placebos. In a 2010 Harvard Medical School study, patients suffering from irritable bowel syndrome were given pills in a bottle labeled "placebo" and at the end of the three-week trial 59 percent of the patients taking the placebos claimed their symptoms had been relieved, far outpacing the 35 percent in the non-treatment group. So, this study suggests that the placebo effect works even when there is no deception involved.[6]

Medieval Pharmacy Splits into Two Traditions

After Galen, pharmacy took two distinct paths, especially during the medieval era. Traditional Galenic pharmacy that had an empirical bent to it continued with a number of classically trained encyclopedists and physicians who kept his tradition alive by translating and compiling his work, including: Oribasius of Pergamum (325–403 CE), Alexander of Tralles (525–605 CE), and Paul of Aegina (625–690 CE). The other path, that began with Pliny and Scribonius Largus, launched into folk medicine with an emphasis on herbalism that advanced the art of the apothecary until the emergence of alkaloid chemistry in the early nineteenth century.[7]

Monastic Medicine and Herb Gardens

An important phenomenon that emerged during the early Middle Ages was the rise of monasteries and convents that preserved and, over time, advanced both strands of pharmacy. The medieval monastery was a self-contained and self-sufficient community of people committed to the work of God, which included a moral obligation to care for the sick. St. Benedict of Nursia (480–554 CE) founded the first of these monasteries at Monte Cassino (in Italy) in 529 CE with the idea of creating a religious community that balanced the need for solitude and reflection with the need to be a member of a community. While Benedict placed the emphasis on prayer and divine intervention in healing patients, over time the monks did work to preserve what was left of the knowledge of antiquity by hand copying important texts, including medical texts. The impetus for preserving ancient medical texts came from an influential Benedictine monk named Marcus Aurelius Cassiodorus (circa 490–585 CE) who urged monks to

> Study with care the compounding of drugs. If you have no knowledge of Greek, you have at hand the *Herbarium* of Dioscorides, who fully described the flowers of the fields and illustrated them with drawings. After that read Hippocrates and Galen . . . and other books dealing with the art of medicine, all of which I have left you on the shelves of the library.

Nearly every monastery had an herb garden and an infirmary. The typical monastic herb garden grew medlar trees, sage, cabbage rose, rosa mundi, pear trees, red peppers, southernwood, rosemary (St. Mary's Rose), St. John's Wort, lemon balm, agrimony, chicory, orpine, rue, spearmint, squill, columbine, tarragon, thyme, parsley, and other herbs. Initially, monastic infirmaries were used to care for sick monks, but subsequently began to treat lay patients. Lay patients were treated in separate buildings known as hospices. These hospices also served as shelters for the poor. Over time, these infirmaries expanded into hospices and then into hospitals. For example, one of Europe's earliest hospitals, the *Hotel Dieu* founded in 1443 in France, literally means "house of God." Despite the edict by the Lateran Council of 1215 CE that banned monks and nuns from treating lay patients, by 1500 CE there were thousands of hospitals throughout Europe equipped with patient wards, pharmacies, and medical libraries.[8]

Medieval Diagnostics

While medieval medical practice was a mixture of Galenic humoral tradition and faith-healing, medieval physicians did try to diagnose patients using the best tools they had at that time. The physician would examine a patient's urine in a clear glass flask, a process known as uroscopy. This was the precursor to modern urinalysis. Based on the color, smell, and taste the physician would be able to offer a diagnosis.[9] The physician would also take the patient's pulse, not to count it, but to examine its rhythms and intensity for any abnormalities. Similar to modern phlebotomy medieval physicians might examine a patient's blood in a process called hematoscopy.[10] A bizarre practice of offering a prognosis was known as the myth of Caladrius, which involved using a bird. If the bird looked at the patient, it meant the patient would live, and conversely, if it looked away, it meant death. As in ancient times astrology and numerology were also employed in offering a prognosis.

The Medical School at Salerno

One of the bright spots during the High Middle Ages was the founding of a medical school at Salerno in the tenth century CE. The seaport town of Salerno already had won renown since Roman times as a health spa. Legend has it that the medical school at Salerno was established by an ecumenical group including: Adale the Arab, Salernus the Latin, Pontus the Greek, and Elinus the Jew.[11] While evidence of its founding remains sketchy, Salerno was an ideal site for the confluence of Greco-Roman and Arab-Islamic medicine, with the Benedictine monastery at Monte Cassino only 70 miles away. Salerno breathed new life into the study of pharmacy and medicine that had not been seen since Galen's time. The medical school attracted some of the finest scholars of the Middle Ages. The first text used by Salerno medical students, entitled *Passionarius Galeni*, was written in about 1050 by Gariopontus, who drew upon the works of Galen, and as a skilled teacher clarified the terminology of the great masters of antiquity. A similar text called *Practica* by Petroncellus was also used.[12] Arab-Islamic medicine was introduced to students at Salerno by a Carthaginian scholar known as Constantinus Africanus (c.1020–1087) who translated Greek, Arab, and Persian medical treatises into Latin for further study. He translated *Antidotarium* written by Sabbatai Ben Abraham also

Donnolo mentions about 120 medicinal plants including: frankincense, opium, origanum, liquid pitch, mint, ivy, hyssop, aloe, black hellebore, ammoniac, aspodel, opoponax, cedar, citron, balsam, artemisia, aristolochia, pennyroyal, wild onion, veronica, genista, tragacanth, rose-laurel, zedoary, olive, ginger, and sesame.

known as Donnolo (913–970) which introduced Arab and Islamic *materia medica* to Europe.[13]

In the late eleventh century, the *Antidotarium magnum* or *Great Antidotary* appeared, describing 485 drug formulas, 200 of which were simple remedies that were prepared for the poor.

Nicholas Praepositus or Nicholas Salernitatus who became the director of the medical school at Salerno in the late eleventh century revised the *Great Antidotary* to make it easier to use. For example, he pared down the list of complex remedies to about 150 and alphabetized them. With this work, he set the standard for drug preparations for centuries by including a table of weights and measures including the now familiar grain, scruple, and drachm. He promoted the use of poly-pharmaceuticals including a formula called *Potio Sancti Pauli* or *Potion of St. Paul*, which was used for epilepsy. One of the most important formulas was for *spongiasomnifera* or anesthetic sponge. The anesthetic sponge's ingredients included opium, henbane, mulberry juice, lettuce, hemlock, mandragora, and ivy. The patient inhaled this from the sponge and was revived later by applying fennel juice to the nostrils. These remedies were often prepared once or twice a year by a *confectionarus* or a *stationarus* who worked under the supervision of a physician in a *statio*, which became the precursor to the pharmacy shop.[14]

Trotula of Salerno and Women's Health

One of the remarkable aspects of life at Salerno in the twelfth century was that women attended classes, and one of its outstanding medical professors was the mysterious Trotula or Dame Trot. She allegedly wrote *De Mulierum Passionibus*, a treatise on obstetrics that advised women about all aspects of child birth. This work remained the definitive work until the sixteenth century. The historical Trotula remains elusive. Some speculate that she was a midwife or perhaps a physician's wife. Others argue that "Trotula" was the name given to describe all midwives in Salerno. She was mentioned in Chaucer's *Canterbury Tales* as "Dame Trot."[15]

One of the key texts that medical students studied at Salerno was the *Regimen sanitatassalernitanum* or *The Salerno Book of Health* that was written in verse with humor and aphorisms conducive to committing to memory. Arnald of Villanova (1235–1315) was credited with revising and expanding it. The book was printed for the first time in 1480 and contained advice on diet and drugs.[16] The medical school at Salerno had an immense impact on pharmacy and medical education during the High Middle Ages as its example inspired the founding of other medical schools throughout Europe. Medical schools and universities opened in Paris (1110), Bologna (1113), Oxford (1167), Cambridge (1209), Montpellier (1181), Padua (1222), Naples (1224), Prague (1347), and Vienna (1365). The medical school at Salerno was closed by Napoleon in 1811.[17]

Hildegard of Bingen

One of the more remarkable figures of the twelfth century was Hildegard of Bingen (1098–1179). Inspired by divine visions as a child, she became a nun and became the abbess of the Benedictine Convent of Rupertsberg near Bingen in modern Germany. She wrote the *Liber Simplicis Medicinae* or *Book of Simple Medicines* which contained nine books that dealt with the curative power of animals, minerals, and plants. She was especially interested in the healing powers associated with bezoar stones. These were kidney or gall stones taken from animals, which she believed had special healing powers. She also wrote a book on the causes, symptoms, and cure of disease called *Liber Compositae Medicinae* that also dealt with psychology. Another of her books was *Liber Operum Simplicis Hominis* that dealt with human anatomy and physiology. Her work revealed that she recognized the significance of the brain and nervous system and understood important elements of the circulatory system. She also speculated about autoimmune diseases. Her advice was sought by kings and traveling pilgrims alike.[18]

The Origins of the European Apothecary

As towns grew in complexity during the twelfth century the need for specialization became apparent and distinct guilds emerged that would separate pharmacy from medicine. For example, in 1178 the Church charter of Cahours, France, implicitly mentioned *apothecarii* with other classes of townspeople. In 1180, the Guild of Pepperers of London specifically referred to apothecaries as members of the guild. Several cities in Europe required apothecaries (sometimes referred to as *especiadors*) to take oaths as in Montpellier, France, in 1180. Other cities that had ordinances pertaining to apothecaries include Marseilles, Avignon, Arles, Melfi, and Venice. Apothecaries often were members of guilds that included other sellers of herbs including: spicers, grocers, perfumers, and pepperers.[19] The guilds operated according to the traditional rules of medieval guilds. They set standards for training apprentices (clerks) and journeyman, regulated membership of the guild, set product standards, and controlled prices for products. They also administered examinations requiring aspiring journeymen who wished to become masters to produce masterpieces. A typical masterpiece project for a journeyman apothecary was to prepare a complex preparation that would satisfy the standards set by the guild. Vestiges of this type of on-the-job training remain a prominent component of a pharmacist's education today in the form of clerkships, rotations, and residencies. For example, today's *Il Nobile Collegio Chimico Farmaceutico Romano* in Italy can trace its heritage back to an apothecaries' guild in the Middle Ages.[21]

> The Anglo-Saxon term *stenc* meant scent, odor, or perfume, giving us the modern English word "stench."[20]

Frederick II and the *Constitutiones*

Between 1231 and 1240 the Holy Roman Emperor Frederick II issued a landmark edict known as the *Constitutiones* that officially separated the practice of pharmacy from medicine. The edict applied to the part of Frederick II's realm known as the Kingdom of the Two Sicilies, but had an impact on the development of pharmacy on the Continent. In order to protect the public, apothecaries' shops (*apotheca*) were subject to inspections by members of the medical faculty of the local medical school known as Collegium medicum, in this case from the medical faculty at Salerno. Apothecaries were obliged by oath to prepare drugs of uniformly good quality. The monarchy set drug prices and limited the number of apothecary shops that could operate. This provision raised both the income and social status of apothecaries at that time. Physicians and apothecaries were forbidden to collude openly or clandestinely in any way to share the profits made from any prescriptions, but in practice both parties bent the rules on occasion. Any violations of the aforementioned provisions could result in the confiscation of the apothecary's property and flagrant violations could even be punishable by death. The *Constitutiones* had an immense impact on the development of pharmacy and medicine on the Continent, but had little effect on the development of pharmacy and medicine on the British Isles and subsequently on its North American colonies.[22]

Medieval Drugs

The Middle Ages was above all else an age of faith and that was reflected in the drugs that were used during that age. The mystical form of medicine that medieval physicians practiced was an amalgam of Galenic humoral theory mixed with faith and folk medicine. Medieval physicians used about 1,000 natural substances including plants, animals, and minerals, often preparing poly-pharmaceuticals. As previously discussed, *Terra Sigillata*, *Hiera Picra*, and *theriac* continued to be used extensively but were supplemented with other uniquely medieval preparations. The doctrine of signatures took on a new significance in that divine providence had marked and imbued various natural substances with great healing power. Compound poly-pharmaceuticals had healing powers, as a result of their preparation, greater than any of their singular ingredients. This idea derived from Galen continued with Avicenna and later formed the basis of homeopathic medicine founded by Samuel Hahnemann in the nineteenth century.

Medieval Drug Remedies

One of the most popular remedies used during the Middle Ages was unicorn horn which was ground up into a powder and was used to treat every condition ranging from infertility to plague. Unicorn powder could be added to any drink and acted as a prophylactic against poisoning. Apothecaries used the horn of oryx, rhinoceros, stag, and especially the tusk of the narwhal. The tusks of narwhals were proudly displayed in apothecary shops and became a symbol of pharmacy. During a visit to a German pharmacy today a person would likely see the words *Einhorn Apotheke* prominently displayed as a reminder of this era.

Another remedy that emerged during the Middle Ages was mandrake. Mandrake was used as a panacea to ingest, and its root was often used as a talisman to ward off illness and to bring good fortune to its wearer. Mandrake root was a woody substance that often resembled a human form and could be carved into various shapes to bestow supernatural powers to its owner. According to medieval legend, since mandrake was imbued with mystical power it had to be harvested carefully. Dogs were often used to uproot the plant and the harvester was advised to carry and play a bugle to ward off the terrible scream the plant would make upon harvesting. Europe's fascination with mandrake lasted until the seventeenth century.[23] In recent times mandrake has regained some its former renown in the Harry Potter book series.

As a result of the Crusades, Europeans broadened their horizon and imported new natural substances from the East. As a result of the contact with the Islamic world, Europeans became interested in the healing properties associated with precious stones and gems. The color, shape, and luster of gems suggested the healing power they might have. Bezoar stones from animal caries of the kidney or gallbladder were reported to have extraordinary healing power and, similar to gems, were carried as amulets to ward off illness.[24] With the rise of the Venetian Republic during the early thirteenth century new natural substances were imported from the East, such as sugar which, due to its cost, was exclusively used in medicines. As trade increased traffic in aloe, benzoin, camphor, cinnamon, cloves, cubebs, ginger, musk, nard, nutmeg, opium, pepper, and rhubarb increased and they were used in the preparation of medicines.

Just when Western Europe appeared to be making strides in advancing medicine and pharmacy through the rise of the universities, a number of extraordinary calamities converged during the late Middle Ages that threatened the very survival of Western civilization. As promising as the thirteenth century was, with the rise of towns and the revival of trade, the fourteenth century proved to some contemporary observers to be the coming of the apocalypse itself. Widespread

By the fourteenth century scented mouth washes became widely used to mask poor dental hygiene. One formula called for a tincture of myrrh dissolved in strong wine that was diluted with an infusion of mint. Tooth picks emerged during this time as well.[25]

peasant revolts, the Hundred Years' War, and the bubonic plague ripped asunder the fabric of what was left of the medieval synthesis. The Great Plague still lingers in modern memory as one of the most terrifying events in human history. From 1347–1351 the plague killed anywhere from a quarter to a third of the European population. The plague struck a population that was utterly unprepared for it. Physicians and apothecaries could not explain the plague and abandoned patients. Similarly, the Church which had been the bedrock that held medieval society together could not explain the plague and clergy abandoned their parishioners to suffer horrible deaths. For example, the Pope abandoned his pastoral duties and secluded himself in a heated room covered in red for a year. Family members abandoned their relations. Corpses piled up with no one to bury them. No institution or individual could offer a satisfactory explanation for this collective tragedy. Consequently, medicine and pharmacy which had made great strides until the plague, lost the trust and respect they had once enjoyed, and it would take centuries to recover their reputations. The plague left an indelible mark on our collective conscience and it continues to trouble us with the prospect that something similar might happen again. The most recent reminder came in the last century with the 1918 flu pandemic that claimed tens of millions of lives worldwide.

Chapter Summary

Medieval physicians and apothecaries explained illness in terms of faith and charity. To be sure, superstition abounded and often hindered the rise of rational medicine, but advances in pharmacy and medicine did occur. The *Constitutiones* recognized pharmacy as separate from medicine and, borrowing the concept from the Islamic world, pharmacy shops and medical schools were established to not only preserve but to advance pharmacy and medical knowledge. Apothecary shops emerged with the rise of medieval towns, and the guild system ensured the competence of their members.

Key Terms

Placebo effect	Myth of Caladrius	Hildegard of Bingen
Benedict of Nursia	Medical School at	bezoar stones
monastic medicine	Salerno	guilds
Lateran Council of 1215	Donnolo	*Constitutiones*
uroscopy	*Great Antidotary*	mandrake
hematoscopy	Trotula	

Chapter in Review

1   Compare and contrast ancient medicine with medieval medicine.
2   Describe the diagnostic techniques a medieval physician would use to examine a patient.
3   Describe the impact the Church had on medieval pharmacy and medicine.
4   Identify the origins of the hospital and trace its evolution into modern times.
5   Describe the significance of the medical school at Salerno.
6   Explain the significance and impact of Frederick II's *Constitutiones* on pharmacy.

Notes

1   George Rosen, *A History of Public Health* (New York: MD Publications, Inc., 1958), 53.
2   A.C.Wootton, *Chronicles of Pharmacy Vol. 1* (1910, Reprint, Boston, MA: Milford House, Inc., 1972), 19. C.J.S. Thompson, *The Mystery and Art of the Apothecary* (London: John Lane the Bodley Head, Ltd., 1929), 78–79. George Bender, *Great Moments in Pharmacy* (Detroit, MI: Northwood Institute Press, 1966), 44–47. Sami K. Hamarneh, "Cosmas and Damian in the Near East: Earliest Extant Monument," *Pharmacy in History*, Vol. 27 (1985), No. 2, 78–83.
3   M. Patricia Donahue, *Nursing: The Finest Art* (2nd ed., New York: Mosby, 1996), 86–89.
4   Suzanne White, "Medicine's Humble Humbug: Four periods in the Understanding of the Placebo," *Pharmacy in History*, Vol. 27 (1985), No. 2, 51–59. Norman Cousins, *Human Options* (New York: Norton, 1981), 207. Quotes two studies conducted by Dr. A.K. Shapiro "A Contribution to the History of the Placebo Effect," *Behavioral Sciences* (1960) and "Attitudes Toward the Use of Placebos in Treatment," *Journal of Nervous and Mental Diseases* (1960).
5   Nathan A. Bostick, Robert Sade, Mark Levine, and Dudley M. Stewart, Jr., "Placebo Use in Clinical Practice: Report of the AMA Council on Ethical and Judicial Affairs," *Journal of Clinical Ethics*, Vol. 19, No. 1 (Spring 2008), 58–61. "Placebos are regularly used by 50 percent of doctors," *St. Louis Post-Dispatch*, October 24, 2008, A14.
6   "Placebo Effect Works Without Deception," *St. Louis Post-Dispatch*, December 23, 2010, A21.
7   C.J.S. Thompson, *The Mystery and Art of the Apothecary*, 68.
8   George Bender, *Great Moments in Pharmacy*, 48–51.
9   Lois N. Magner, *A History of Medicine* (New York: Marcel Dekker, Inc., 1992), 103.
10 Roy Porter, *The Greatest Benefit to Mankind: A Medical History of Humanity* (New York: Norton, 1997), 115.
11 A.C.Wootton, *Chronicles of Pharmacy Vol. 1*, 115–116.
12 David L. Cowen and William H. Helfand, *Pharmacy: An Illustrated History* (New York: Harry Abrams, 1990), 50–51. C.G. Cumiston, *The History of Medicine* (New York: Dorset, 1987), 214.
13 C.J.S. Thompson, *The Mystery and Art of the Apothecary*, 128.
14 David L. Cowen and William H. Helfand, *Pharmacy*, 51.

15 Jeanne Achterberg, *Woman as Healer* (Boston, MA: Shambala Publications, Inc., 1990), 48–49. C.G. Cumiston, *The History of Medicine*, 218–219.
16 C.J.S. Thompson, *The Mystery and Art of the Apothecary*, 133.
17 Glenn Sonnedecker, Comp., *Kremers and Urdang's History of Pharmacy* (4th ed., Madison, WI: American Institute of Pharmacy, 1976), 33.
18 Jeanne Achterberg, *Woman as Healer*, 54–57.
19 Charles Lawall, *The Curious Lore of Drugs and Medicine: Four Thousand Years of Pharmacy* (Garden City, NY: Garden City Publishing, Inc., 1927), 136.
20 Sally Pointer, *The Artifice of Beauty* (Gloucester: Sutton Publishing, Ltd., 2005), 59.
21 David L. Cowen and William H. Helfand, *Pharmacy*, 53.
22 Sally Pointer, *The Artifice of Beauty*, 55. James Grier, *A History of Pharmacy* (London: The Pharmaceutical Press, 1937), 43. George Bender, *Great Moments in Pharmacy*, 60–63.
23 David L. Cowen and William H. Helfand, *Pharmacy*, 56–57.
24 John M. Riddle, "Lithotherapy in the Middle Ages: Lapidaries Considered as Medical Texts," *Pharmacy in History*, Vol. 12 (1970), No. 2.
25 Sally Pointer, *The Artifice of Beauty*, 83.

6 Medieval Arab and Islamic Pharmacy

How did the Golden Age of Science in the Islamic World advance pharmacy and medicine?

The era known as *Islamic medicine* or *Arab medicine* describes the type of medicine practiced in medieval Islamic civilization that was written in Arabic, which was the official language of the Islamic civilization. While Islam originated in Arab civilization, it spread rapidly in the seventh century into a great empire that extended from India in the East to what would become Spain in the West. The point here is that the Islamic world includes many civilizations and people who share the Islamic faith, but are not ethnically of Arab descent. In fact, Islamic medicine includes traces of ancient Greek, Hebrew, Indian, Egyptian, and Persian medicine, in addition to the discoveries made by its own practitioners during the Middle Ages.

Even before Muhammad's time (570–632 CE) there was contact among the Arab, Greek, Jewish, Roman, Persian, and Indian worlds. In the fifth century CE, Nestorian Christians, followers of the recently excommunicated Patriarch Nestor, were expelled from Constantinople and founded a teaching hospital at Jundishapur in Persia. Jundishapur also became a haven for Greek philosophers who had fled from the Academy in Athens when the Byzantine Emperor Justinian closed it, bringing Greco-Roman knowledge with them.[1] The Arab world, under the leadership of enlightened caliphs such as al-Mansur and Haroun al-Raschid in the eighth century, saw the value of classical ideas and preserved this vast body of knowledge, especially during its Golden Age from the seventh through the thirteenth centuries.

The Birth of Alchemy

One of the greatest scientific advancements that occurred during this Golden Age was the advancement of chemistry, an indispensable component of pharmacy. Chemistry derives from the Egyptian word *khem* which means "black" referring to the rich black soil deposited on the banks of the Nile River. Sometimes chemistry has been referred to as the black art. It also has been referred to as the "hermetic art," after the Egyptian god Hermes Trismegistrus (thrice great).[2] The first mention of chemistry appeared in a third-century CE Roman manuscript by Sextus Julius Africanus. The Arabs added the prefix *al* to the Egyptian word *khem* yielding the English word "alchemy" which, during the Middle Ages, meant searching for the ability to transform common substances into precious metals.

Geber

During the eighth century CE, Jabir ibn-Hayyan (*c.*721–815) known in English as "Geber," became a legendary figure in the development of alchemy. Some modern historians of Islamic pharmacy consider Jabir as the father of modern medicinal chemistry. More than 200 treatises have been attributed to him with some appearing after his death, leading historians to believe that his work was expanded upon by his students. In addition to improving the mortar and pestle, he developed tongs, stills, crucibles, and stoves. He became the "Father of Alchemy" due to his descriptions of distillation, sublimation, and calcination. By the use of his chemical processes, Jabir brought the whole of the mineral kingdom into the realm of therapeutics and chemical pharmacy. Legend has it that he discovered sulfuric acid (oil of vitriol), nitric acid, and silver nitrate. In Islamic civilization alchemy and pharmacy were separate fields of inquiry; they nonetheless exerted great influence on each other.[3]

Among the many experiments he conducted in his laboratory, Jabir accidentally discovered alcohol. Legend has it that he tasted this substance and when his lab assistant found him, he was talking nonsensically or speaking "gibberish."

Mesue Senior

Another important figure in the history of Islamic pharmacy was Mesue Senior (*c.*777–837) or John the Damascene who was the son of an apothecary at Jundishapur. Although he was a Nestorian Christian he went on to become the head of the medical school at Baghdad and was the attending physician to the six caliphs. His drug formulary became the model for the first London Pharmacopoeia. He rejected the harsh purgatives used in Greco-Roman medicine and favored the reintroduction of ancient Egyptian and Ayurvedic laxatives including senna, cassia fistula, tamarind, and jujube. Along with his students Mesue also translated the entire *Corpus Hippocraticum* into Arabic, an achievement that insured this important work would be preserved for posterity.[4] A point of particular note was the focus on preventive medicine through diet and traditional dietetics that served as the major theme of the *Corpus*

Hippocraticum. This served as the foundation in the evolution of advanced nutrition in medieval al-Andalus.

Al-Kindi

A contemporary of Mesue was the ninth-century CE scholar al-Kindi (*c*.796–874) or Alkindus. Al-Kindi served as a major influence on the English botanical pharmacist/alchemist Roger Bacon (1214–1298), who laid the foundation for developing the scientific method in the West. In his book *De Gradibus*, al-Kindi showed how mathematics could be used to advance pharmacy.[5] He developed mathematical scales to measure the strength of drugs. The methods developed by al-Kindi led to the ability to quantify the degree of hotness or coldness in a drug in the pharmacology of humors. This methodology is still in use today, e.g. the modern Scoville unit (SU) which measures the degree of hotness (heat) in the pepper family ranging from zero SU in the green bell pepper to upwards of 250,000 SU in the Thai chili pepper to 16,000 SU in pure capsaicin itself, the compound that confers heat to chili peppers.

Albucasis

Another scholar was the Spaniard Abu al-Qasim al-Zahrawi (936–1013) or Albucasis, who wrote *Kitab al-Tasrif* or *The Method of Medicine* which contains a treatise called "Liber Servitoris," an influential work on medicinal chemistry. Albucasis was known for inventing many medications, including one to treat the common cold that he called *Muthallaathat*, which was made from camphor, musk, and honey that resembles today's Vicks Vapor Rub. He also developed nasal sprays, hand lotions, and mouth washes. He also invented medical plaster and adhesive bandages. Albucasis' work filtered into the West and became a standard text in several European medical schools for centuries.[6]

Rhazes

One of the greatest physicians of all time Abu Bakr Muhammad ibn Zakariya Razi (865–925) or Rhazes was born in Persia and spent his early years working for an apothecary while studying music. After visiting a hospital he decided to devote his life to the study of medicine. Well read in many subjects and well informed about Greek, Indian, and Persian medicine, Rhazes wrote over 200 treatises. His most enduring work was a 14-volume medical encyclopedia known as the *al-Hawifi'l-tibb* or *Continens* (*Comprehensive Book of Medicine*). Similar to Hippocrates' work, this work was probably begun by Rhazes and augmented by several generations of his students. He taught at the medical school in Baghdad and attracted an international following including students from as far away as China. He introduced case studies, separate wards in hospitals, and the use of pills in administering medication.

He was the first physician to distinguish measles from smallpox through his keen clinical observations and case studies. Rhazes divided medicine into three areas including public health, preventive medicine, and the treatment of diseases. He came tantalizingly close to understanding germ theory when asked to select a site to build a new hospital. His experience told him that wounds healed faster in certain places than others and he put this to the test in an experiment. He hung pieces of meat at various sites around Baghdad. He selected the site where the meat decomposed the least to build the hospital. Although he earned handsome fees from his elite patients, Rhazes donated his small fortune to help the needy and died destitute.[7]

Avicenna

Another Persian physician/philosopher who was destined to make his mark on pharmacy history was Abu Ali Hussain ibn Abdullah ibn Sina (980–1037) or Avicenna. A child prodigy, he had memorized the entire *Qu'ran* by the age of ten and had graduated from medical school at 16. He wrote over 200 medical treatises; his most significant was *Kitab al-Qanun* or *Canon of Medicine* a comprehensive medical encyclopedia that deals with nearly every aspect of medicine. Two of the volumes deal with pharmacy, a volume on simples, and a volume on compounds including a cancer compound called *hindiba*. Avicenna's lasting contribution to pharmacy was the gilding and silvering of pills to make them easier to swallow.[8] The *Canon* was one of the first books to place medicine on the path toward evidence-based medicine, calling for the experimental use and testing of drugs for their efficacy. The pharmaceutical concepts of the *Canon* guided Western medicine until the seventeenth century, but the *Canon* remains in print and still plays an integral part of medicine practiced in Central Asia.[9]

Avicenna wrote a brief book dedicated solely to the therapeutics of cardiovascular diseases. One of the simple drugs listed in this specialized formulary has recently been found to have beta-blocker activity. Beta-blockers are a class of modern anti-hypertensive drugs pioneered in the 1960s by Sir James Whyte Black (1924–2010). Black won the Nobel Prize for Medicine and Physiology in 1988.

Ibn Zuhr

While the works of Rhazes and Avicenna garnered praise from many, in Islamic medicine there was room for skepticism and debate. The first who openly challenged the emerging conventional wisdom of the great medical masters was Ibn Zuhr (1091–1162) or Avenzoar who opposed the use of astrology and mysticism that had influenced pharmacy and medicine for

Figure 6.1 Avicenna (*c*.980–1037 CE) a medieval Persian apothecary and physician wrote one of the most important and enduring medical texts in world history, the *Canon of Medicine.*

Source: Painting by Robert Thom, reproduced courtesy of the American Pharmacists Association Foundation.

centuries. Ibn Zuhr was born in Seville and grew up as a polymath amid the rich culture of medieval al-Andalus. His book *Kitab al-Taisir* (*Book for the Study of Therapeutics and Diet*) dealt with specific disease states and challenged some of Galen's teachings. He derived his knowledge from testing surgical procedures on animals and conducting autopsies on humans. In another book entitled *Kitab-al-Agdiya* (*Book on Food*) he described not only the culinary and nutritional aspects of food, but also discussed the preventive and medicinal properties of foods. The science of dietetics which developed in Islamic al-Andalus, due in part to Ibn Zuhr, has proven to have tangible health benefits which come down to us today as the general dietary pattern called the Mediterranean diet.[10]

Averroes

A fellow Spanish-born philosophical skeptic and colleague was Averroes who, among other things, questioned the immortality of the soul leading his works to be banned among both Muslims and Christians alike. Abu'l-Walid Muhammad ibn Ahmad ibn Muhammad ibn Rushd (1126–1198) or Averroes was better known to history as a philosopher, but in his medical writing he raised doubts about the theoretical medicine advocated in Avicenna's *Qanon* and promoted his own work on pharmacy called the *al-Kulliyat* or *The Book of General Principles*.[11] Dietetics in the *Colliget* included recommendations about the quantity and frequency of meals, the order in which the various foods should be eaten (e.g. salad or soup as the first course to aid in digestion) and modification of the diet based on age and state of health. Together Avenzoar's and Averroes' works were translated into Hebrew and Latin and became collectively known as the *Colliget*, a work of independent medical thought that influenced the West.[12]

Maimonides

Averroes' best known student was a Spanish-born Jew known as Rabbi Moshe ben Maimum (1135–1204) or Maimonides. Forced to flee his native Cordova due to religious persecution, Maimonides sought refuge in Morocco, finally settling in Cairo where he became the attending physician to Saladin. Serving in this exalted position he became fascinated by *theriac* and *mithridatium* and wrote a treatise about toxicology and the treatment of poisonous bites. While he wrote a number of medical treatises his most famous was the "Oath and Prayer of Maimonides" that contains the immortal and often quoted phrase: "May I never see in the patient anything but a fellow creature in pain."[13] In this oath we see the confluence of ethics and medicine.

The Oath and Prayer of Maimonides

Thy Eternal Providence has appointed me to watch over the life and health of Thy creatures. May the love for my art actuate me at all times; may neither avarice, nor miserliness, nor the thirst for glory, nor for a great reputation engage my mind; for the enemies of Truth and Philanthropy could easily deceive me, and make me forgetful of my lofty aim of doing good to Thy children.

May I never see in the patient anything but a fellow creature in pain. Grant me the strength, time, and opportunity always to correct what I have acquired, always to extend its domain; for knowledge is immense and the spirit of man can extend infinitely to enrich itself daily with new requirements. Today he can discover his errors of yesterday and tomorrow he may obtain a new light on what he thinks himself sure of today.

O God, Thou hast appointed me to watch over the life and death of Thy creatures; here I am ready for my vocation.

And now I turn unto my calling.
O stand by me, my God, in this truly important task;
Grant me success! For ___
Without Thy loving counsel and support,
Man can avail but naught.
And for Thy creatures,
O, grant _____
That neither greed for gain, nor thirst for fame, nor vain ambition,
May interfere with my activity.
For these, I know are enemies of Truth and Love
of men,
And might beguile one in profession
From furthering the welfare of Thy creatures.
O strengthen me.
Grant energy unto both body and soul
That I might e'er unhindered ready be
To mitigate the woes,
Sustain and help
The rich and poor, the good and bad, enemy and friend.
O let me e'er behold in the afflicted and suffering,
Only the human being.[14]

One of the distinguishing features of Islamic medicine during the Middle Ages was the introduction of a new form of pharmacy literature in the form of formularies that served as guides to apothecaries in the preparation of drug recipes and for their medicinal uses. Arranged in alphabetical order for easy referencing, these formularies served as practical guides to apothecaries. Saburibn-Sahl's *al-Aqrabadhin al-Kabir*, which appeared in the mid-ninth century was the first of several important formularies. This was followed by ibn-'AbdRabbih's *al-Dukkan* or *The Apothecary Shop* in the tenth century, which discussed dosage forms for various medicaments. Al-Biruni's (973–1050) *Kitab al-Saydanah fi al Tibb* or *Book of Pharmacy in the Healing Art* described over 1,000 simples but, more importantly, described the role of the apothecary and their duties and functions. This formulary showed that the apothecary or *sayadilah* had arrived as an integral part of Islamic medicine. A more extensive formulary by Ibn-al-Baytar (*c.*1188–1248) of Spain entitled *Kitab al-Jami* described 800 botanical drugs, 145 mineral drugs, and 130 animal drugs. In addition to formularies directed at apothecary shops there were formularies written for hospital pharmacies, the most significant of which was the *Dusturbimaristani* or *Hospital Formulary* written by the Jewish physician al-Salidibn Abi'l-Bajan who practiced at the Nasiri Hospital in Cairo in the thirteenth century. Since the ninth century, apothecaries filled prescriptions written by physicians at Islamic hospitals such as the Nuri Hospital in Damascus. Apothecaries often held staff appointments at hospitals such as the hospital in Marrakesh during the twelfth century.[15]

Ibn al-Baytar: Medical Ethnobotanist

Ibn al-Baytar was perhaps the most accomplished botanical apothecary of Islamic science. He became the leading expert of his time on the medical and botanical aspects of horticulture and agriculture. He wrote a work containing 1,000 simple drugs (most of them food plants), 300 of which he discovered himself. He later became the botanist and chief attending physician to the caliph in Cairo. Al-Baytar studied plants on his extensive travels throughout al-Andalus, North Africa, and the Middle East becoming the model for medical ethnobotanists, who study how different cultures use medicinal plants.

The Handbook for the Apothecary Shop

Perhaps the greatest formulary written during this era began as a treatise written by a concerned father who wanted to pass along his experience as an apothecary in Cairo to his son. The treatise entitled the *Minhaj al-Dukkanwa Dustur al-'yan* or *Handbook for the Apothecary Shop* was written by a Jewish apothecary named Abu al-Muna Kohen al-'Attar in 1259. In addition to being a valuable formulary on how to prepare medicines, the *Handbook* advised apothecaries to keep their shops clean, well-stocked, neat, and attractive. Apothecaries were advised to "keep profits moderate." Similarly, al-'Attar urged apothecaries to be friendly, honest, thoughtful, slow to anger, modest, and patient.[16]

With the advancement of scientific learning the apothecaries flourished in the Islamic world. The advances and expansion of alchemy and the drug trade prompted the need for trained specialists who could prepare the increasingly complex prescriptions independently from physicians. Under these conditions apothecaries became recognized as specialists with a distinct identity and function. By the order of the caliph, Abu Coreisch Isa al Szandalani opened the first pharmacy shop in Baghdad around 762. These pharmacies were open-front stores with a service counter lined with glass and clay jars. Behind the counter there were scales, mortars, and other compounding instruments.[17] The *dakakin al-sayadilah* or educated apothecary shop owner eventually became differentiated from the traditional *sayadilah* or *attarin* (spicers) who often had no formal training.[18]

Drugs and Spices that the Islamic World Introduced to the West

Camphor, senna, Chinese rhubarb, musk, cassia, tamarinds, manna, saffron, benzoin, Indian hemp, aconite, datura, sandalwood, cinnamon, cardamoms, nutmegs, cloves, cubebs, turmeric, galangal, ginger, amomum, and cane sugar.[19]

Chapter Summary

At a time when Western Europe was reeling from the aftermath of the collapse of the Roman Empire, another civilization in the Middle East was embarking upon a Golden Age of scientific and mathematical discovery. The ancient Arab people began as nomadic tribesmen who roamed the Arabian Peninsula, herding livestock and passing down a rich oral cultural heritage. In the seventh century CE, the prophet Muhammad began his ministry and founded one of the world's great religions, Islam. Islam unified the Arab people and gave them a written literary language, i.e. Arabic, which revolutionized many fields of knowledge. It is fair to say that without the advancements made during this Golden Age of Islamic culture, pharmacy and medical education as we know it would not be possible. Words that have Arab origins that have passed down into English reflect the influence of this Golden Age and include alchemy, algebra, alcohol, elixir, sugar, jujube, and syrup to name a few. In fact, the Arab word *dowa* came down through the Romance languages as *doga* or *droga* yielding the modern English word *drug*.[20] The pioneering efforts of Geber in alchemy and of Rhazes, Avicenna, and others in medicine, who learned from the Greco-Roman masters, synthesized this tradition with the best of Egyptian, Indian, and Persian medicine to create the best medicine of its time. This Golden Age produced the first full service pharmacy shops with educated and licensed apothecaries. This tradition also promoted the use of pills to administer medication and invented the gilding and silvering of pills. A Cairo apothecary, Abu al-Muna Kohen al-Attar, wrote the *Minhaj* which gave apothecaries their own set of ethical standards of practice. By the high Middle Ages, Islamic medicine spread to Europe and prompted the founding of Western Europe's first medical school at Salerno, which in turn gave rise to the founding of the great universities of Europe that would help set the stage for a new era of learning in the Renaissance.[21]

Key Terms

Alchemy	Rhazes	Maimonides
Geber	Avicenna	Ibn-al-Baytar
Mesue Senior	gilding and silvering	*Minhaj sayadilah*
Alkindus	Avenzoar	
Albucasis	Averroes	

Chapter in Review

1 Understand the distinction between Arab culture and the Islamic faith.
2 Account for the success of the Golden Age of Science in the Islamic world during the eighth–thirteenth centuries.
3 Describe the birth of alchemy and its significance in the evolution of pharmacy.
4 Explain the contributions of Islamic apothecaries, physicians, and philosophers to the evolution of pharmacy.
5 Account for the influence Islamic pharmacy and medicine had on the development of pharmacy and medicine in Western Europe.

Notes

1 A.C. Wootton, *Chronicles of Pharmacy Vol. 1* (1910, Reprint, Tuckahoe, NY: Milford House, 1972), 101–102. David L. Cowen and William H. Helfand, *Pharmacy: An Illustrated History* (New York: Harry Abrams, 1990), 39. James Grier, *A History of Pharmacy* (London: The Pharmaceutical Press, 1937), 32.
2 Ibid. (Grier), 123. Charles Lawall, *The Curious Lore of Drugs and Medicine: Four Thousand Years of Pharmacy* (Garden City, NY: Garden City Publishing, Inc., 1927), 95.
3 Ibid., 94, 102.
4 Ibid., 102–103. Fielding H. Garrison, *An Introduction to the History of Medicine* (Philadelphia: W.B. Saunders Company, 1929), 128.
5 Charles Lawall, *The Curious Lore of Drugs and Medicine*, 104.
6 Ibid., 105.
7 George Bender, *Great Moments in Medicine* (Detroit, MI: Northwood Institute Press, 1965), 63–67.
8 Joseph B. Sprowls, Jr. and Harold M. Beal, *American Pharmacy: An Introduction to Pharmaceutical Technics and Dosage Forms* (Philadelphia: J.B. Lippincott Company, 1966), 10.
9 George Bender, *Great Moments in Pharmacy* (Detroit, MI: Northwood institute Press, 1966), 56–59.
10 A.C. Wootton, *Chronicles of Pharmacy, Vol. 1*, 110–111.
11 Charles Lawall, *The Curious Lore of Drugs*, 109–110.
12 Ibid., 110. Roy Porter, *The Greatest Benefit to Mankind: A Medical History of Humanity* (New York: Norton, 1997), 100.

13 Ibid., 112.
14 Charles Lawall, *The Curious Lore of Drugs*, 112–113.
15 Ibid., 101–102.
16 David L. Cowen and William H. Helfand, *Pharmacy*, 49. Roy Porter, *The Greatest Benefit to Mankind*, 102.
17 C.J.S. Thompson, *The Mystery and Art of the Apothecary* (London: John the Bodley Head, Ltd., 1929), 81.
18 David L. Cowen and William H. Helfand, *Pharmacy*, 46.
19 James Grier, *A History of Pharmacy*, 33.
20 Ibid., 33.
21 The author would like to publicly thank Dr. Cedric Baker, Pharm.D. of Mercer University for his careful reading of the draft chapter, his contributions to this chapter, and for offering his expert comments about its contents.

7 Pharmacy During the European Renaissance and Early Modern Era

Why do historians argue that the Renaissance marks the beginning of the modern era and what implications did this have on pharmacy and medicine?

The Great Plague claimed many victims and not all of them were people. The type of thinking and institutions that had dominated the Middle Ages failed to explain the utter devastation the plague had wrought and had failed to protect what proved to be a very vulnerable population. To be sure, for generations the plague continued to affect those who survived its wrath but the survivors began seeking new answers to the eternal questions about life and its meaning. Ironically, many sought answers by studying the wisdom of the ancient world with an eye toward applying the best of these ideas to their own time. They were looking back in order to move ahead. Over time, a series of subtle changes in values, attitudes, and technologies occurred that would eventually shape the modern world. For example, two remarkable men, the great Florentine poet Petrarch and the writer Giovanni Boccaccio, survived the devastation of the plague, and had already demonstrated in their writing a shift from the sacred or liturgical language of the medieval era to thinking and writing in secular terms. Boccaccio's *Decameron*, a book that chronicles the devastation of the plague, remains one of the most incisive books about the plague and how it affected one town.[1] The medieval values of piety, obedience, poverty, and humility were slowly giving way to modern values of money, pleasure, and power.

Along with the new types of thinking came new technologies that transformed the world. The magnetic compass, the Caravel, maps, and moveable metal type literally opened new worlds, and, for pharmacy, dramatically expanded the Old World's *materia medica*. The Renaissance or "rebirth" in the late fifteenth and sixteenth centuries proved to be a landmark time in world history for the discovery of new ideas and places. Niccolo Machiavelli, a Florentine diplomat, shocked the world with his insider's account of modern power politics that established the blueprints that led to the absolutist nation-state. Similarly, Baldassare Castiglione in his *Book of the Courtier* discussed how an educated person should go about his business in this new world of power politics and commerce. Nicholas Copernicus, a Polish astronomer who worked as a clerk for the Church, speculated that the earth revolved around the sun, a notion that gave birth to modern science. Andreas Vesalius (1514–1564), the son of a Belgian apothecary, diplomatically disproved Galen's explanation of human anatomy and launched a new field of biology. Martin Luther, a German Augustinian monk, would take on the Western world's most powerful institution, the Church, and by insisting on the power of individual conscience, began his own church and changed Christianity forever. Christopher Columbus, a sea captain in search of a shortcut to Asia, discovered an entirely New World and all of a sudden new ideas and new commodities (a good number of which were spices and drugs) flooded into Europe. By the power of guns, germs, and steel the Europeans, for better or worse, transformed the earth, including medicine and pharmacy. Johann Gutenberg (1400–1468), a German printer from Mainz, discovered moveable metal type and launched a communications revolution that allowed all of the aforementioned pioneers to publish their ideas and circulate them to a wide audience. The sharing of powerful ideas over space and time was here to stay and shaped the modern world.[2]

Paracelsus

The free-thinking spirit of the Renaissance created the ideal conditions for one of the most controversial figures in the history of pharmacy to appear. Philippus Aureolus Theophrastus Bombastus von Hohenheim (1493–1541), better known to history as "Paracelsus," transformed medicine and pharmacy forever. Rarely accused of modesty, Paracelsus reportedly took this name meaning "super Celsus" or "above Celsus" to demonstrate the superiority of his ideas over those of the ancients, in this case of the great Roman encyclopedist, Aulus Cornelius Celsus, whose *De re Medicina* was one of the first books printed using the printing press in 1478. He was the son of a Swiss physician who had an insatiable curiosity and free spirit. He traveled widely throughout Europe and beyond. Some accounts have him visiting Russia and China. From his own accounts, Paracelsus enjoyed learning about the healing arts from people from all walks of life including barber-surgeons, midwives, and alchemists, among others. From his travels, he developed a deep abiding respect for folk medicine, which did not endear him to his colleagues. He reportedly studied at the medical schools in Ferrara, Basel, and Vienna with conflicting accounts as to whether he ever completed any of these programs.[3]

By many accounts, Paracelsus wound up as an army surgeon in the service of the Holy Roman Emperor Charles V, successfully serving in the Empire's war against France. Part of Paracelsus' success as an army surgeon was his insistence that wounds be kept clean and probed as little as possible. His reputation as a healer of the first order spread rapidly. In a stroke of

Figure 7.1 Paracelsus (1493–1541) was a self-educated itinerant surgeon who rejected Galenic humoral theory in favor of seeking pharmaceutical cures for illness. Paracelsus remains one of the most controversial figures in the history of medicine.

Source: Painting by Robert Thom. Collection of the University of Michigan Health System, Gift of Pfizer Inc. UMHS.11.

good fortune, Paracelsus was called to the bedside of a wealthy publisher named Frobenius who suffered from an infected leg wound that would not heal. Under Paracelsus' care Frobenius' leg healed and he arranged for Paracelsus to receive a teaching appointment at the nearby University of Basel Medical School. It did not take long for Paracelsus to run afoul of the administration and faculty at the university, as he delivered his lectures in German and not the customary Latin. Furthermore, Paracelsus called upon his students to bring their textbooks to be burned in a public bonfire on campus as a symbolic rejection of the theories of the great medical masters such as Galen and Avicenna. The students adored him, but the administration and faculty forced him out in 1528 by restricting student access to him and his classes. He was banned from Basel over a dispute over a medical fee and his highly critical inspections of local apothecary shops. For the rest of his life Paracelsus wandered from town to town, most often as a *persona non grata*. His father died in 1537 leaving him some property which supported Paracelsus' work until his own mysterious death in 1541 in Salzburg.

Speculation about the causes of his death, range from severe alcoholism to mercury poisoning to cancer.[4]

From his travels and the practical experience he gained by observing those involved in the healing arts of his time, Paracelsus rejected Galenic humoral theory, an anathema for this time in history. Medical students who questioned Galen's teachings would fail their medical exams and physicians who did so would be ostracized from the medical community, a fate Paracelsus wore with distinction. By all accounts, Paracelsus was a skilled alchemist who developed a number of effective remedies composed of a few ingredients, unlike the complicated poly-pharmaceuticals prepared by his contemporaries. As an alchemist he viewed the human body as a vast chemical laboratory that was guided by a vital force he called the *archaeus*. For Paracelsus, disease was a function of nature that could be mitigated or cured by the appropriate natural or chemical substance. Paracelsus replaced Galenic humoral theory with three primary principles of the human body: sulfur (combustibility), mercury (volatility or liquidity), and salt (stability). Any imbalance of one of these principles in the body caused illness. This might appear similar to Galenic humoral theory, but the similarity ends when one understands that Paracelsus' theory focused treatment on a specific part of the body, not the entire body as in the Galenic system. In this way diseases could be identified and treated, not merely the symptoms of disease. Paracelsus' assumption was that since nature caused illness, there had to be some substance in nature that could cure it. His idea that specific drugs could cure specific diseases opened up a new way of looking at the relationship between diseases and their treatments. For this reason he has been called the "father of iatrochemistry."[5]

As a skilled alchemist, Paracelsus called upon his peers to abandon their search for gold and instead use their talent to find new drugs that could cure disease and improve the human condition. Paracelsus experimented with a number of substances including powdered tin, zinc, zinc salts, mercury, copper, lead, antimony (which he called stribium), arsenic, and iron.[6] From these he formed compounds. For example, his use of mercury compounds rather than mercury itself to treat syphilis yielded positive results. He also prepared a tincture of opium he called laudanum. While Paracelsus was a man ahead of his time in advancing the idea that drugs held the key to curing human illness, he was also a product of his time. Paracelsus was also deeply mystical in his beliefs. His belief in astrology and the role it played in healing harked back to ancient times. This led him to revive the idea of the doctrine of signatures. He believed nature placed hints in plants and minerals that indicated their utility in healing. For example, he used hellebore root to treat epilepsy, but insisted that it must be harvested on a Friday morning during the waning of the moon.[7]

One of Paracelsus' most bizarre experiments involved the creation of human life itself. He claimed that by mixing human sperm, blood, and horse manure buried for 40 days he had produced a fully functioning miniature human being he called a "homunculus." Alchemists all over Europe attempted to replicate his experiment, but none of them could produce a homunculus for all to see. Pressured by the public to see his homunculus, Paracelsus claimed that he had killed it because it did not have a human soul, as only God could bestow a soul.

For all of his talent, Paracelsus remained a wanderer literally and symbolically all of his life, in search of better ways to heal human suffering, primarily through drug therapy. By rejecting Galen's teachings, Paracelsus pointed the way out of traditional medical thought toward a new paradigm. In terms of pharmacy, he moved it from a concentration on botany toward chemistry. While his own life was brief and tumultuous, he pioneered concepts in alchemy for his followers to pursue, which added greatly toward advancing pharmacy.

Ambrose Pare

Another important figure who experimented and changed his thinking about medical treatment during this era was the French surgeon Ambrose Pare (1510–1590). As a battlefield surgeon Pare followed the conventional wisdom of his time and treated gunshot wounds by pouring boiling oil into them. The thinking at that time was that gunshot victims were poisoned by the gunpowder and therefore by cauterizing the wound with boiling oil of elder mixed with *theriac* surgeons were purifying it. After treating many soldiers with gunshot wounds in the wake of an especially ferocious battle, Pare ran out of boiling oil and was forced to improvise with the supplies he had available. He prepared a salve consisting of egg yolks, oil of roses, and turpentine and applied it to the soldiers' wounds. He spent a very anxious night, carefully monitoring all of his patients and to his astonishment the patients who had received his improvised salve were improving, whereas the soldiers who had received cauterization with boiling oil suffered from fevers, swelling, and pain. From this point forward, Pare vowed to abandon the conventional treatment and to pursue more effective and less painful ways of treating patients. For example, the conventional surgical procedure for amputation was to staunch the bleeding at the site of the amputation with hot cautery irons. In 1552, he reintroduced the use of ligatures that had been abandoned after antiquity to tie off the blood vessels, and abandoned painful cautery. Once again, in the true Renaissance spirit Pare looked back to the ancients to advance medicine. Similar to Paracelsus, Pare faced heavy criticism from several prominent professors of medicine, but responded to their objections by writing *The Apology and Treatise* in which he defended his work. Pare also became known as the "father of prosthetics" for his design and development of artificial body parts to try to make peoples' lives whole again.[8]

Maggot Therapy for the Treatment of Wounds

The great French battlefield surgeon, Ambrose Pare, noted after the Battle of St. Quentin in 1557 that soldiers whose wounds had become infected with maggots healed faster and with fewer complications than soldiers whose wounds had not been infected with maggots. Maggots had been used since that time, but the first surgeon to scientifically study them was Dr. William S. Baer (1872–1931) from Johns Hopkins University. With the use of antibiotics in the 1940s, maggot therapy fell out of favor, but is still used in some circles as a treatment for slow healing wounds and ulcers.[9]

Andreas Libavius

In addition to Paracelsus and Pare there were a number of alchemists, apothecaries, botanists, and physicians who advanced the study of pharmacy in the sixteenth century. While not exactly a follower of Paracelsus, the German physician and alchemist Andreas Libavius (1540–1616) agreed that alchemy should be directed toward finding drugs to cure diseases. Toward this end, he wrote what might be called the first chemistry textbook, *Alchemia*, in 1595. He rejected Paracelsus' mysticism and argued that chemistry must be based on observation and experimentation. He discovered stannous chloride which, for many years, was known as *Spiritus Fumans Libavii* or Libavius' fuming liquor. He also experimented with crude methods of chemical analysis, testing various minerals.[10]

Nicholas Houel

Another interesting sixteenth-century figure was the French apothecary Nicholas Houel (1520–1584) who had become wealthy on the proceeds he earned from his shop. He used his wealth to fund a House of Christian Charity that became a school for young orphans. The orphanage had a chapel, a school, a pharmacy, a herb garden, and a hospital. Eventually, the hospital became the *Hotel des Invalides*. The school taught the children reading, writing, religion, and the art of the apothecary. For a time the school came under the control of the Roman Catholic Church, but in 1622 was given back to the Society of Apothecaries. In 1777, the school became a College of Pharmacy.[11]

Sir Theodore Turquet de Mayerne

A follower of Paracelsus, Sir Theodore Turquet de Mayerne (1573–1655) was born in Geneva and studied medicine at Heidelberg and later Montpellier.

He taught anatomy and pharmacy at the University of Paris where the faculty members were staunch Galenists and drove him out. Turquet rose to become an attending physician to King Henry IV, but due to long-standing professional jealousies complicated by religious persecution, he was expelled from France. He fled to England where he became the attending physician to King James I, Charles I, and Charles II. His medical advice also was sought by Oliver Cromwell.[12] In England, he went by the name de Mayerne. Along with Francis Bacon, Henry Atkins (a prominent apothecary), and Gideon de Laune (apothecary to the Queen), he played a key role in establishing the "Master, Wardens and Society of the Art and Mystery of the Apothecaries of the City of London" in 1617. For the first time in the English-speaking world, apothecaries had their own organization separate and distinct from grocers and spicers.[13]

The London Pharmacopoeia

Building on this success, de Mayerne, a member of the College of Physicians, also helped produce the first *London Pharmacopoeia* in 1618 and wrote the preface to it that reads:

> [W]e venerate the age-old learning of the ancients and for this reason we have placed their remedies at the beginning, but on the other hand, we neither reject nor spurn the new subsidiary medicines of the more recent chemists and we have conceded to them a place and corner in the rear so they might be a servant to the dogmatic medicine, and thus they might act as auxiliaries.[14]

The *Pharmacopoeia Londinensis* had sections on plants, metals, and minerals, but also had age-old recipes including ingredients such as bowels of mole, mummy powder, hog's grease, raspings from an unburied human skull (to treat gout), among others. The *Pharmacopoeia* also listed recent chemical medicaments including crocus of antimony, vitrified antimony, mercury of life (butter of antimony), mercuric sulfate, and calomel, an acknowledgment of the growing significance of chemical compounds for medical treatment. De Mayerne himself used calomel as a purgative in his practice.[15]

Jean Van Baptiste Van Helmont

Perhaps the most ardent follower of Paracelsus was the Flemish aristocrat Jean Baptiste Van Helmont (1579?–1644). He studied at the University of Louvain and began teaching at 17. A man of great compassion, he devoted his life to the poor which was best exemplified by his treatment of plague victims in Antwerp in 1605. He fell ill himself, became disgusted with the heavy doses of purgatives used to treat him, and devoted himself to the study of alchemy to find better cures for illness. He discovered carbon dioxide and called

When Van Helmont studied at the University of Louvain he refused to accept any formal degree. He believed that formal academic degrees only led to hubris and pomposity. Due to his skill in effecting cures through his medicines, he was once called before the Inquisition![16]

it *gas sylvetre* coining the term "gas." He discerned that gas was a separate entity from the atmosphere. His main work *Ortusmedicinae* or *The Garden of Medicine* did not appear until after his death. He was referred to as "Doctor Opiatus" because he prescribed opium to his patients so frequently. Complementing the work of Van Helmont, was François de la Boe Sylvius (1614–1672) a French alchemist who developed the doctrine of iatrochemistry by which disorders of the body could be classified by being either acidic or alkaline in nature. An excess of either acid or alkaline in the body could be corrected by employing a chemical to rebalance the chemicals in the body. Iatrochemistry or medical chemistry was a fusion of alchemy, medicine, and chemistry.[17]

John Rudolf Glauber

Perhaps the last of the German alchemists, John Rudolph Glauber (1604–1668) was born in Carlstadt and from his youth became fascinated by the natural world. Suffering from a stomach ailment, Glauber cured himself by drinking mineral water and set out to analyze the chemical in the water that cured him. He identified the chemical as sulfate of soda, a purgative salt, which would be known to history as Glauber's Salts. He conducted further experiments by distilling sulfuric acid with sea salt, which yielded a spirit of salt known today as hydrochloric acid. He noted the gas that emerged from the salt, but curiously did not identify it as chlorine. His experiments with ammonia, sulfate of copper, and various chlorides advanced the art of wine-making. His best known work was *De Natura Salium* which, as with all of his works, bears a Latin title, but was written in German.[18]

Rembert Dodoens

While mainstream apothecaries, physicians, and alchemists were advancing pharmacy, this era also proved to be a productive time for advancing herbalism. Rembert Dodoens (1517–1611?) was a Flemish physician and botanist. He studied medicine at the University of Leuven and became a court physician to the Hapsburg king, Rudolf II. Rudolf II was an ardent supporter of medicine, alchemy, astrology, and science and his capital city of Prague became a haven for all types of research presided over by luminaries such as Tycho Brahe (1546–1601) and Johannes Kepler (1571–1630). Dodoens' *Cruydtboeck*, which was published in 1554, discussed six groups of plants, including a chapter on medicinal plants that resembled a pharmacopoeia. Dodoens' herbal book would later serve as the basis for herbal books in Britain.

In Britain botanical studies and herbalism flourished during the Renaissance. The first attempt was the *Grete Herball* which was published in 1526 by an anonymous author. This work was followed by two books *Libellus de re herbaria* (1538) and *The Names of Herbes* (1551) by the Cambridge-educated physician and botanist William Turner (1520?–1568). In 1597, the English surgeon and herbalist John Gerard (1545–1612?) published the *Herball* or *Generall Historie of Plantes*. This book relied heavily on Rembert Dodoen's work and Gerard added his own comments about each plant as well as 1,800 woodcut illustrations. By all accounts, Gerard was a devoted gardener who kept a world-class herb garden at Holborn.[19]

Nicholas Culpeper

The most prominent and controversial herbalist of this time was Nicholas Culpeper (1616–1654). The son of a clergyman, Culpeper studied medicine at Cambridge and was disinherited by his family for doing so, but seemed destined to become a prominent physician when tragedy struck. He fell in love and planned to be married, but his fiancée was struck by lightning and killed. Overwhelmed by grief, Culpeper quit his studies and took on an apprenticeship with an apothecary. He eventually married the daughter of a wealthy merchant which allowed him to open up his own shop just outside of London. This allowed Culpeper to practice the type of medicine he had always wanted without the direct interference of the London authorities. He was a man of deep convictions and compassion for the poor and wanted everyone to have equal access to medical treatment. Culpeper believed in using as many locally grown herbs as possible, and by keeping his costs down, he was able to treat many people for free. Rejecting much of the traditional medical treatment he had learned about at Cambridge, he scoffed at what he saw as pompous, self-serving, and useless medical practices such as uroscopy and bleeding. Instead, he relied on herbal treatments and astrology, which in retrospect might have done some good and certainly did less harm than the harsh bleeding and purging that traditional medicine offered. Although he was a remarkably skilled apothecary, he soon ran afoul of the authorities because he was treating patients and prescribing drugs on his own.[20]

Undaunted by the pressure on him, Culpeper translated the *London Pharmacopoeia* in 1649 from the Latin into English to make it more accessible to apothecaries and others who wished to read it. The translation became

Figure 7.2 Nicholas Culpeper (1616–1654) began his career studying medicine, but rejected it in favor of searching for herbal cures for illness and published one of the most influential books in the history of pharmacy; the *Complete Herbal*.

known as *Physicall Directory* and almost immediately drew the wrath of the College of Physicians who were alarmed that this interloper was making medicine more transparent, revealing their guarded trade secrets for all to read. In 1653, Culpeper published his most famous and most lasting work, *The English Physician Enlarged*, which became known as *Culpeper's Herbal*. The book has withstood the test of time, going through at least 40 editions, and is still in print. The book describes over 500 plants, 369 of which were native to England. He also described in English many plants and herbs from the New World. The herbal also described foxglove, the plant from which digitalis would later be made, which was used to treat heart ailments.[21]

Culpeper died of tuberculosis in 1654, a champion of apothecaries and a medical maverick to the end. The popularity of Culpeper's work made it easier for medical writers to publish their work in the vernacular rather than in Latin. Culpeper's work also had a dramatic impact upon the development of medicine and pharmacy in the English-speaking colonies of the New World. Culpeper's books went to the New World and the spirit of transparency and self-help they taught took root, especially in North America.[22]

Drugs From the New World

One of the most exciting developments of this era was the discovery and introduction of new herbs from the New World. The most significant in terms of effectiveness were cinchona and ipecacuanha, but others included coca, guaiacum, tobacco, and sarsaparilla. The discovery and introduction of cinchona into the European *materia medica* remain shrouded by myth and legend. Cinchona was native to South America, especially abundant in Peru, which came under Spanish control by the mid-sixteenth century. The bark of the cinchona plant contains medically active alkaloids including quinine, which is effective in treating the symptoms of malaria. Due to its effectiveness against fevers, the natives called it *quina-quina* or the "bark of barks."[23] It was used by the Incas and other natives to South America and was introduced to Jesuit missionaries who brought it to Europe in the mid-seventeenth century. For this reason, cinchona often was referred to as "Jesuit's bark." The Italian botanist Pietro Castelli wrote about it in the 1640s. Cinchona became part of the *London Pharmacopoeia* in 1677. Carolus Linnaeus (1707–1778), the Swedish botanist, named it after the Countess of Cinchona, the wife of a viceroy of Peru, who was successfully treated for malaria with it and brought it back to Europe upon her return to Spain in 1640.[24] Cinchona was a major breakthrough in treating the symptoms of malaria and was one of the few drugs that actually worked for many centuries. In the early nineteenth century, the French chemists Pierre-Joseph Pelletier (1788–1842) and Joseph-Bienaime Caventou (1795–1877) studied the chemical composition of cinchona and in 1820 successfully isolated two distinct alkaloids they named "quinine" and "cinchonicine."[25] Later, in the nineteenth century, cinchona would play a key role in helping Dr. Samuel Hahnemann develop the principles of homeopathic medicine.

Ipecacuanha

Another plant from South America that also had an impact on the development of purgatives and emetics was *Cephalis ipecacuanha* or ipecacuanha. Ipecacuanha was first mentioned in 1625 in a Portuguese friar's account of Brazil which describes it as a remedy for "bloody flux," better known today as dysentery. It could also be used to induce vomiting or as an expectorant to control coughs. Called *ipecaya* by the natives, it was first brought to Paris by a traveler named Le Gras in 1672. Ipecacuanha came to notice in Europe in 1686 when it was introduced by Jean Adrien Helvetius who sold it to King Louis XIV as a treatment for diarrhea.[26] Today syrup of ipecac can usually be found in homes where there are small children or pets, as a precaution, to be used to induce vomiting in case of accidental poisoning.

The Foxglove Plant and Digitalis

Another useful drug of the time was foxglove, which was native to Europe and, over time, became a widely used and effective treatment for dropsy or the swelling of tissue due to heart disease. In the twelfth century, it was known to the Anglo-Saxons. Fuchius of Tubingen mentioned the plant in his *Plantum Omnium Nomenclature* in 1541 and gave it the name *digitalis*.[27] It was used to treat an excess of phlegm and was included in the herbals of William Turner and John Gerard. In 1650, foxglove was included in the *London Pharmacopoeia*. The discovery that showed the efficacy of digitalis in treating dropsy was made in 1775 by Dr. William Withering (1741–1799).[28] Withering, of Birmingham, England, had heard about and experimented with a traditional folk remedy for dropsy that local people had used in tea for decades.[29] He used it in powder form or sometimes as an infusion. He had documented cases where he showed that *Digitalis purpurea* was an effective treatment for dropsy and was the first to recognize its value in treating heart ailments. In 1785, he published his findings in *An Account of the Foxglove and Some of its Medical Uses etc; With Practical Remarks on Dropsy and Other Diseases*. Withering recognized that digitalis was potentially toxic to the heart and called for it to be administered in gradual doses, a practice that continues in effective drug therapy today.[31]

> The foxglove plant was known as "bloody fingers" or "dead men's bells" in Scotland. In German the plant is called finger-hut (finger hood) or thimble.[30]

Guaiacum

While cinchona and ipecacuanha have stood the test of time for their efficacy other drugs imported from the New World, which upon discovery had held great promise, have faded from modern memory. One of these drugs was guaiacum which was a New World cure for syphilis (then called *Morbus Gallicus*), a disease that had ravaged Europe since the Age of Exploration. Prior to the discovery of guaiacum, the customary treatment for syphilis and the sores it produced was to administer salves composed of quicksilver (mercury salts), pig's fat, and two irritant herbs—spurge, and oil of stavesacre. A poem entitled, *Syphililides, sive Morbi Gallici* or *Syphilis or on the Gallic Disease*, by the Italian physician and apothecary Hieronymus Frascatorus (1478–1553), described in detail the ordeal of this treatment.[32] The ointment was applied to the skin, except to the head and area over the heart, and the patient was swathed in bed covering and left to undergo profuse sweating. This regimen was continued for ten days until massive salivation in the mouth led to profuse drooling which was taken as a sign the poison was leaving the body and the cure was taking effect. What actually occurred were the symptoms of mercury poisoning with severe ulceration of the mouth and jaw area, sometimes complicated by tremors and paralysis.[33]

When Spaniards living in the Caribbean observed natives using draughts from the guaiac tree (*Guaiacum officianale*) to treat what appeared to be symptoms of syphilis, but was probably yaws, they were eager to try the cure. Hoping to capitalize on the new cure, Gonsalvo Ferand shipped the first guaiacum wood to Spain in 1508 and earned a fortune from it. The guaiacum trade attracted the attention of the leading banking dynasty of Europe, the Fugger family, who obtained a monopoly on its trade. Testimonials to the new drug's success spread rapidly throughout Europe. The German humanist Ulrich von Hutten (1488–1523) wrote about his ordeal with syphilis, recounting how he suffered through ten regimens of treatment with quicksilver only to find relief from one regimen of treatment with guaiacum in 1519. Ironically, von Hutten suffered from a relapse and died in 1523, dealing a serious blow to the claims of guaiacum being a miracle cure. The arduous regimen of treatment associated with guaiacum, consisting of fasting, confinement, no alcohol, and celibacy, proved to be too much for many patients to endure and they abandoned the treatment. The treatment regimen weakened many patients beyond their ability to recover from it. Also, the Caribbean natives who used guaiacum also used a rigorous course of sweat baths in sweat lodges, which probably had more to do with killing the syphilis spirochaete than the guaiacum. Guaiacum treatment had largely been abandoned when it was revived briefly by the Dutch physician Hermann Boerhaave (1668–1738) in the eighteenth century.[34]

Tobacco

Almost from their initial encounter with it in the New World the Europeans were fascinated by the natives' use of tobacco. Jean Nicot, an ambassador to Lisbon for Francis II of France, introduced it to Europe in 1559, although some sources credit Sir Walter Raleigh with this honor.[35] Ironically, Nicot gave his name to the plant called *Nicotiana tabacum* as well as the name of its chief alkaloid nicotine. Its praises were sung by many who viewed it as an antidote to all poisons and its ability to heal wounds better than St. John's Wort.[36] For better or worse, tobacco products became a mainstay in retail sales for many pharmacies and its example serves as a cautionary tale about the human tendency to be fascinated by novelty without applying a healthy dose of skepticism.

From *Apothecaire* to *Pharmacien*: How Apothecaries Became Pharmacists

From the fourteenth to the seventeenth century, the French *apothecaire* enjoyed steady progress in social prestige. In addition to preparing and administering drug preparations according to pharmacopoeia standards, some apothecaries also prepared and administered enemas or clysters, usually for wealthy patients. This practice became a medical fad from the fifteenth to the eighteenth century and made fortunes for many apothecaries. For example, King Louis XIII of France reportedly received at least 312 enemas within a single year. To some contemporary observers, this practice bordered on the ridiculous and became a prime target for satirists. The most talented satirist of his time was Jean Baptiste Poquelin (1622–1673) better known by his pen name "Molière." In his play *Le maladie imaginaire*, Molière lambasted the physicians and apothecaries who administered these enemas for such exorbitant fees. Molière showed the giant syringe that was administered in a ridiculously sanctimonious manner to patients. Cartoons and caricatures appeared that soon turned French public opinion against this practice, prompting apothecaries to opt for a better image by referring to themselves as *pharmacien*.[37]

The Pharmacy Lexicon in the United States

A few centuries later across the sea in North America, apothecaries were developing a distinct American identity and were in search of an appellation that captured their essence. Soon after the American Pharmaceutical Association (APhA) was formed in 1852, its leadership began to develop a lexicon to describe what its members did and how they operated. The APhA Constitution defined its membership as "pharmaceutists" and "druggists," defining the former as retail apothecaries and the latter as wholesale dealers.[38] In 1853, there was a proposal to rename pharmaceutists as "pharmacians," but it failed. It was not until 1866 that Edward Parrish convinced his fellow APhA members at the annual meeting that pharmaceutists should be known as pharmacists.[39] In 1926, this led the APhA to promote the term "pharmacy" over the commonly used term "drugstore." After a study conducted by Glenn Sonnedecker in 1948, the APhA tried to get pharmacists to move away from a retail model to a more professional model by urging its members to "dispense prescriptions" rather than "sell prescriptions." In 1959, the APhA convinced telephone book publishers to delete the category "drugstore" in favor of "pharmacy." In the same year "retail" and "neighborhood" pharmacy gave way to "community pharmacy." In 1954, the term "pharmaceutical care" was first coined to describe how pharmacists interacted with patients. As clinical pharmacy grew in the 1980s, Charles D. Hepler used the term to describe the care patients received from clinical pharmacists. In 1990, Linda Strand and Charles Hepler took the term to its logical conclusion in describing how clinical pharmacists were assuming the responsibility for their patients' drug therapy and overall health outcomes. Medication therapy management (MTM) services came into the lexicon as a result of federal legislation in 2000 with the full support of the APhA.[40]

Chapter Summary

In the wake of the Great Plague, many of the medieval institutions that held power were challenged by the new scholarship. The use of moveable metal type created a communications revolution, an explosion of information that resonates to our own time. The religious language that had been used since civilization began was being replaced with a secular and, later, scientific language to explain how the world worked. The discovery of the "New World" ushered in a flood of new plants and substances that rapidly expanded the European *materia medica*. Pioneers such as Paracelsus challenged the medical establishment and laid the foundation for the Scientific Revolution which profoundly shaped the future of pharmacy.

Key Terms

Paracelsus	Jean Baptiste Van	digitalis
Ambrose Pare	Helmont	William Withering
prosthetics	John Rudolph Glauber	dropsy
Nicolas Houel	Rembert Dodoens	guaiacum
Theodore de Mayerne	Nicholas Culpepper	calomel
Pharmacopoeia	*The English Physician*	Jean Nicot
Londinensis (The London	cinchona (quinine)	Molière
Pharmacopoeia)	ipecacuanha (ipecac)	

Chapter in Review

1 Explain how the Renaissance era transformed pharmacy.
2 Explain the complex place Paracelsus holds in the history of pharmacy and medicine.
3 Evaluate the respective careers of Paracelsus, Ambrose Pare, and Nicholas Culpeper.
4 Explain and assess the impact that plants from the New World had on pharmacy and medicine.
5 Account for how apothecaries became pharmacists.

Notes

1 Giovanni Boccaccio, *Tales From the Decameron*, trans. Richard Aldington (Chicago, IL: Puritan Publishing Company, Inc.), 1930.
2 A.C. Wootton, *Chronicles of Pharmacy, Vol. 1*(Boston, MA: Reprint, Milford House, 1972), 236.

3 Roy Porter, *The Greatest Benefit to Mankind: A Medical History of Humanity* (New York: Norton, 1997), 201.

4 Barbara Griggs, *Green Pharmacy: The History and Evolution of Western Herbal Medicine* (Rochester, VT: Healing Arts Press, 1997), 43.

5 Glenn Sonnedecker, Comp., *Kremers and Urdang's History of Pharmacy* (4th ed., Madison: American Institute of the History of Pharmacy, 1976), 41.

6 Howard W. Haggard, *Devils, Drugs, and Doctors* (New York: Harper and Row, 1929), 345–348.

7 Roy Porter, *The Greatest Benefit to Mankind*, 203. Charles Lawall, *The Curious Lore of Drugs and Medicine: Four Thousand Years of Pharmacy* (Garden City, NY: Garden City Publishing, Inc., 1927), 247. Bruce T. Moran, "The Herbarius of Paracelsus," *Pharmacy in History*, Vol. 35 (1993), No. 3, 99–127.

8 George Bender, *Great Moments in Medicine* (Detroit, MI: Northwood Institute Press, 1965), 92–98.

9 Milton Wainwright, "Maggot therapy: A Backwater in the Fight Against Bacterial Infection," *Pharmacy in History*, Vol. 30 (1988), No. 1, 19–26.

10 Charles Lawall, *The Curious Lore of Drugs and Medicine*, 250.

11 Ibid., 250.

12 Barbara Griggs, *Green Pharmacy*, 70–71.

13 George Bender, *Great Moments in Pharmacy* (Detroit, MI: Northwood Institute Press, 1966), 68–71.

14 Barbara Griggs, *Green Pharmacy*, 72.

15 A. C. Wootton, *Chronicles of Pharmacy, Vol. 1*, 255–257.

16 Ibid., 259.

17 Glenn Sonnedecker, *Kremers and Urdang's History of Pharmacy*, 43.

18 A.C. Wootton, *Chronicles of Pharmacy, Vol. 1*, 261–264.

19 Inge N. Dobells, ed., *Magic and Medicine of Plants* (Pleasantville, NY: Reader's Digest Association, Inc., 1986), 60–61.

20 Charles Lawall, *The Curious Lore of Drugs and Medicine*, 229.

21 Barbara Griggs, *Green Pharmacy*, 94.

22 Ibid., 95. Nicholas Culpeper, *The English Physician*, reprint edited by Michael Flannery (Tuscaloosa: University of Alabama Press, 2007), 1–110.

23 C.J.S. Thompson, *The Mystery and Art of the Apothecary* (London: John Lane the Bodley Head, Ltd., 1929), 228.

24 Ibid., 229–230.

25 George Bender, *Great Moments in Pharmacy*, 100–103.

26 A.C. Wootton, *Chronicles of Pharmacy, Vol. 2* (Reprint, Boston, MA: Milford House, 1972), 114–115. C.J.S. Thompson, *The Mystery and Art of the Apothecary*, 239.

27 C.J.S. Thompson, *The Mystery and Art of the Apothecary*, 243.

28 The term "dropsy" to describe edema or an accumulation of fluid in a body cavity dates back to at least the time of Hippocrates. Paul of Aegina (625–690) described dropsy as a malfunction of the liver. Richard Bright (1789–1858) in 1827 was the first to distinguish between cardiac and renal dropsy.

29 Jeanne Achterberg, *Woman as Healer* (Boston, MA: Shambala Publications, Ltd., 1991), 107.

30 Charles Lawall, *The Curious Lore of Drugs and Medicines*, 109.

31 Barbara Griggs, *Green Pharmacy*, 34. Charles Lawall, *The Curious Lore of Drugs and Medicines*, 251.

32 Ibid. (Griggs), 36–37.
33 Ibid., 37–40.
34 Ibid., 96.
35 Charles Lawall, *The Curious Lore of Drugs and Medicines*, 232.
36 David L. Cowen and William H. Helfand, *Pharmacy: An Illustrated History* (New York: Harry Abrams, 1990), 103–104.
37 Gregory J. Higby, "From Compounding to Caring: An Abridged History of American Pharmacy." In *Pharmaceutical Care*, edited by Calvin H. Knowlton and Richard P. Penna (Bethesda, MD: American Society of Health-System Pharmacists, 2003), 23.
38 Dennis B. Worthen, *Heroes of Pharmacy* (Washington, DC: American Pharmacists Association, 2008), 154–155.
39 George Griffenhagen, ed., *150 Years of Caring* (Washington, DC: American Pharmaceutical Association, 2002), 173–174.
40 Robert M. Elenbaas and Dennis B. Worthen, *Clinical Pharmacy in the United States: Transformation of a Profession* (Lenexa, KS: ACCP, 2009), 114–117.

8 Eighteenth- and Early Nineteenth-Century Pharmacy

From Apothecary to Chemist to Pharmacist

How did pharmacy contribute to the birth of chemistry and how did chemistry influence the future course of pharmacy?

Despite the remarkable discoveries in science and medicine, such as Isaac Newton's mechanical universe and William Harvey's pulmonary circulation of the blood, medical practice still relied on variations of Galen's humoral theory. In the seventeenth century, researchers went from asking "why" to "how" and discovered new paradigms that would transform the ways in which people saw the world and the universe beyond. Even though scientific research advanced new paradigms such as the pulmonary circulation of the blood, medical practice stagnated with the traditional practices of uroscopy and bleeding patients to collect a fee for these services. For example, although Harvey (1578–1657) and Marcello Malpighi (1628–1694) explained the pulmonary circulation of the blood, there was little the average physician could do with this discovery and, thus, without antiseptic surgery, antibiotics, and general anesthesia, open heart surgery could not be performed successfully until the 1960s.[1]

This lag between the advancement of medical theory and the stagnation of medical practice hindered early eighteenth-century medicine. Due to the advances of chemistry in the seventeenth century by the likes of apothecaries including Nicaise le Febvre (1610–1674) and Nicolas Lemery (1645–1715), who wrote influential chemistry textbooks, pharmacy advanced more rapidly during this time than medicine did. This is not a surprising development, since apothecaries already had a long history of preparing medicines with botanical ingredients and were uniquely positioned to experiment and take advantage of the advancements in chemistry. Le Febvre had served as the court apothecary to Louis XIV and later to Charles II of England. Le Febvre wrote *Traite de chymie theorique et pratique (Treatise on the Theory and Practice of Chemistry)*, which was published as *Cours de chymie (Course on Chemistry)* in its fifth edition. The book became quite popular and was translated into several languages.[2] By contrast, Lemery was a self-taught chemist who opened a pharmacy in Paris and taught chemistry there. His lectures drew students from several countries and he became a minor celebrity. He stayed in London at the invitation of King Charles II and was invited to lecture at the University of Berlin. He based his books *Universal Pharmacopoeia* and *Dictionary of Simple Drugs* on his life's work and thereby influenced generations of apothecary/ chemists.[3]

Apothecaries Advance Iatrochemistry

Building upon iatrochemistry, apothecaries/chemists explored fermentation and the relationship between respiration and combustion. The early stages of this gas chemistry had its roots in the seventeenth century with Sylvius' *Disputationes Medicae* published in 1663.[4] Etienne François Geoffrey (1672–1731) exemplified the apothecary/chemist of this era. He was the son of an apothecary and studied pharmacy at Montpellier and returned to earn a degree in medicine. For his doctorate in medicine he submitted three prescient theses; whether all diseases have one origin and therefore might be cured by one drug, whether physicians must also be chemists, and whether human beings are descended from earthworms. He taught pharmacy and medicine for many years at the College of France and was honored by being made a Fellow of the Royal Society of London. He worked on iron, vitriol, fermentation, mineral waters, and wrote an influential essay entitled, *The Superstitions Concerning the Philosopher's Stone*, which argued that chemists ought to find cures for diseases rather than trying to turn dirt into gold.[5]

Rouelle and Lavoisier

Guillaume François Rouelle (1703–1770) served his apothecary apprenticeship in a shop once owned by Nicolas Lemery and later opened his own shop, in which he gave lectures on chemistry. Antoine Lavoisier was said to have attended some of these lectures. Rouelle declined the chance to serve as the king's apothecary so he could devote more time to research. He classified salts into the categories of acid, neutral, and base. He experimented in alkaloid chemistry and came very close to discovering it.[6] Rouelle's student Antoine Laurent Lavoisier (1743–1794) transformed modern chemistry by isolating and naming oxygen as well as developing the language associated with modern chemistry in his seminal work *Traite Elementaire de Chimie* published in 1789. Tragically, Lavoisier ran afoul of the political authorities in revolutionary France for his pre-revolution involvement with tax collection and was executed by guillotine.[7] Prior to Lavoisier, other chemists had recognized that the atmosphere was composed of many gases and they proceeded to identify them. The English chemist Henry Cavendish (1731–1810) had identified hydrogen in 1766 calling it "inflammable air" and Daniel Rutherford (1749–1819) a physician, chemist, and botanist from Edinburgh, Scotland identified nitrogen in 1772.[8]

Phlogiston Theory

Still, it would be Lavoisier's work that would be instrumental in advancing modern chemistry in disproving phlogiston theory. A German physician, Johann Joachim Becher (1635–1682), and his student Georg Ernest Stahl (1659–1734) argued that combustible substances must have a material they termed "phlogiston" that is transferred from the burning substance to the air during combustion. Thus, phlogiston theory was used to explain the phenomena of the burning and rusting of substances. This led to the terms "phlogisticated air" that applied to nitrogen and "dephlogisticated air" that applied to oxygen. Lavoisier, borrowing from the work of the great apothecary/chemist Karl Wilhelm Scheele and Joseph Priestly, was able to show that the "calx" of a metal weighed more than the original metal, dealing a serious blow to phlogiston theory. Nonetheless, advocates of phlogiston theory continued their fight into the nineteenth century.[9]

Joseph Priestly

One defender of phlogiston theory was the free thinking British Unitarian minister-turned-chemist Joseph Priestly (1733–1804). A talented linguist and philosopher, Priestly became a close friend of Benjamin Franklin (1706–1790) who sparked Priestly's interest in science. Priestly began experimenting with carbon dioxide, resulting in the publication of essays entitled *Experiments and Observations on Different Kinds of Air* and *Dephlogisticated Air* that appeared in 1775 and 1776, respectively. He had isolated dephlogisticated air, which would later be named oxygen, by heating a metal oxide. Having had his laboratory ransacked in Britain due to religious persecution, Priestly arrived in the United States in 1794 and published a spirited defense of phlogiston theory in an 1800 essay entitled *The Doctrine of Phlogiston Established*. Priestly died in Northumberland, Pennsylvania in 1804.[10]

Karl Wilhelm Scheele

Karl Wilhelm Scheele (1742–1786) was born into a merchant family in Stralsund (Pomerania) and went on to make great discoveries in chemistry while working all of his life as an apothecary. His career began at age 14 in Gothenburg as an apprentice to local apothecary, Martin Andreas Bauch, where he worked until 1765 when the owner sold the business. He moved on to Stockholm and Upsala, winding up in Koping where he bought his own apothecary shop in 1776. Financially secure and free to experiment, Scheele was able to pursue his love of learning and research. He conducted thousands of experiments in which he discovered oxygen (in 1773, a year before Priestly did, although his work was not published in time due to printing delays), chlorine, prussic acid, tartaric acid, phosphoric acid, tungsten, uric acid, benzoic acid, glycerin, and nitroglycerin. His work *On Fire and*

Air, which explained the findings of his work with gases, was published in 1777. Legend has it that he died from inhaling hydrocyanic acid during an experiment in his laboratory. Scheele accepted phlogiston theory to his dying day and never challenged it. A compilation of his work was published after his death, called *Opuscula Chemica et Physica*. While his work was appreciated during his life, its full scope only came to be recognized after his death for providing the foundation for the field of organic chemistry.[11]

Louis Nicolas Vanquelin

Another apothecary/chemist/educator of this era was Louis Nicolas Vanquelin (1763–1829) whose discoveries paved the way for alkaloid chemistry. His laboratory became a practical school of chemistry in which he investigated the chemical composition of belladonna, cinchona, ipecacuanha, and other herbs. He also discovered the chemical elements chromium and beryllium. Vanquelin separated an impure substance he called nicotine from tobacco in 1809. In 1812, Vanquelin extracted daphine from mezereon root describing its alkaline nature. Vanquelin also served as the founding director of the School of Pharmacy of Paris from 1803–1829, influencing several generations of pharmacists.[12]

Friedrich Wilhelm Adam Serturner and the Discovery of Alkaloid Chemistry

Although the seeds of alkaloid chemistry had been sown with the work of Scheele and Vanquelin, it would take the work of a young German apothecary from Hannover, Friedrich Wilhelm Adam Serturner (1783–1841), to isolate the painkilling substance from the opium plant, calling it *morphium* or morphine. Since ancient times humanity has dreamed of a substance that would relieve unrelenting pain and now here it was. Serturner began experimenting with opium in 1803 and published the results of his experiments that yielded meconic acid in the *Journal der Pharmazie* in 1805. He followed with a more detailed paper published in 1806, in which he described the narcotic principle he called *principium somniferum*. The substance *morphine* was given its name in 1817, after the Greek god Morpheus, who was the god of dreams. Serturner's morphine was alkaline and represented a new class of organic bases that formed salts with either organic or inorganic acids.[13]

Serturner's discovery was met with skepticism and envy from his critics who claimed to have discovered morphine first. Louis Nicolas Vanquelin, Armand Seguin (1765–1865), and Jean-François Derosne (1774–1846) cast doubt on his findings, and Seguin accused Serturner of plagiarism. In 1803, Derosne and Seguin had isolated a substance from opium but could not completely identify it. Despite this controversy, in 1831 the *Institut de France* awarded the *Prix Monthyon* (Monthyon Prize) of 10,000 francs to Serturner for recognizing the alkaline nature of morphine and for advancing medicine. Serturner had

Figure 8.1 F.W.A. Serturner (1783–1841), a German chemist from Hannover, who pioneered alkaloidal chemistry and isolated morphine from the opium plant. His work opened up the development of modern synthetic drugs.

Source: Painting by Robert Thom, reproduced courtesy of the American Pharmacists Association Foundation.

established a new field of chemical study, which the German pharmacist Karl Mesissner called "alkaloid" in 1818. The historical significance of·Serturner's work should not be underestimated because the process of extracting the basic substances from natural herbs led to the development of the pharmaceutical industry, which will be the subject of a forthcoming chapter. For the first time in the history of pharmacy the active ingredient of a plant could be isolated and concentrated into chemical form which had tremendous implications regarding purity, drug standardization, and consistency of dosage.[14]

Opium

Since ancient times, there have been few plants that have had such a profound impact on the course of human history. The white Indian poppy plant *Papaver somniferum* produces fleshy seed capsules from which the juice is dried, yielding opium. Highly addictive, opium has been used medicinally as an effective pain reliever and later was abused as a recreational drug, causing millions of people to become addicted. In the eighteenth century, the British East India Company imported opium into China and used it to trade for silk and precious metals and jewels. By 1839, the British had illegally imported nearly 40,000 chests of opium into China. The Chinese government issued warnings to the British to desist the importation of opium, and when the British refused, the Chinese officials destroyed 20,000 opium chests.

Queen Victoria's Real Secret

The British retaliated militarily, precipitating the first of two Opium Wars. The First Opium War lasted from 1840–1842, in which China lost Hong Kong and five other Chinese ports. As a result of the Second Opium War, China lost more of its sovereignty and the opium trade was legalized, opening the door to widespread addiction.

Advancements in alkaloid chemistry have been a curse and a blessing. Opium's most addictive alkaloids have produced morphine and heroin in the nineteenth century. The heroin trade has become a worldwide problem that has been in the spotlight due to the recent war in Afghanistan.[15]

Joseph Bienaime Caventou and Joseph Pelletier

Building upon Serturner's work in advancing alkaloid chemistry the French pharmacists Joseph Bienaime Caventou (1795–1877) and Joseph Pelletier (1788–1842) worked together as a research team to isolate quinine from the cinchona plant in 1820. Both men worked as pharmacists in Paris while pursuing their passion for alkaloid chemistry. In addition to his discoveries with Pelletier of strychnine, brucine, quinine, and cinchonine, Caventou also coined the term "chlorophyll" that describes the green pigment in the leaves of plants. Pelletier, either alone or in collaboration with others, was a brilliant chemist who isolated emetine from ipecacuanha, strychnine and brucine from nux vomica, and quinine from cinchona. He also discovered narceine, an alkaloid of opium. Both of these men were pure researchers who offered their discoveries to the world without taking out patents on them. For example, Pelletier began manufacturing his preparation of quinine sulfate to treat the symptoms of malaria without a patent. His preparation had widespread appeal, and the Farr and Kunzi Company began manufacturing and distributing quinine in Philadelphia in 1822. In 1823, Rosengarten and Sons also began manufacturing quinine sulfate in Philadelphia. The industrial revolution had come to pharmacy.[16]

The work of Serturner, Caventou, and Pelletier inspired further research in alkaloid chemistry. Pierre Robiquet (1780–1840) studied under Fourcroy and Vanqueline in Paris when he was conscripted into one of Napoleon's armies. After his military service, he opened his own pharmacy and began to manufacture some drugs for sale to other pharmacies. He isolated asparagin when he studied under Vanqueline. Later in 1832, he extracted codeine from the opium plant. The German pharmacist-chemist Phillip Laurence Geiger (1785–1836) discovered conine and, with the German chemist L. Hesse, discovered aconitine, atropine, colchicine, daturine, and hyoscyamine. By the end of the nineteenth century, pharmacy had undergone a complete transformation from herbal drugs to powerful pharmaceuticals, including morphine, codeine, quinine, cocaine, cochicine, ephedrine, atropine, papaverine, reserpine, digoxin, and digitoxin.[17]

Chapter Summary

While the debates in the early eighteenth century centered on phlogiston theory, the latter part of this same century saw an astounding series of discoveries that yielded morphine and quinine, two chemicals that went on to revolutionize drug therapy. The advent of alkaloid chemistry also provided the impetus behind the beginning of the pharmaceutical industry.

Key Terms

Nicaise Le Febvre

Nicholas Lemery

Etienne François Geoffrey

Guillaume François Rouelle

Antoine Laurent Lavoisier

phlogiston theory

Joseph Priestly

Karl Wilhelm Scheele

Louis Nicolas Vanquelin

Friedrich Wilhelm Adam Serturner

morphine

Joseph Bienaime Caventou

Joseph Pelletier

quinine

Pierre Robiquet

Chapter in Review

1   Explain the contribution apothecaries made to advancing alchemy into modern chemistry.

2   Identify and describe the contributions eighteenth- and early nineteenth-century apothecary/chemists made in advancing pharmacy.

3 Explain the historical significance of alkaloid chemistry in the history of pharmacy.

Notes

1 Roberto Margotta, *The History of Medicine* (New York: Smithmark Publishers, 1996), 94.
2 Glenn Sonnedecker, comp., *Kremers and Urdang's History of Pharmacy* (4th ed., Madison: AIHP, 1976), 354.
3 A.C. Wootton, *Chronicles of Pharmacy, Vol. 1* (Reprint, Boston, MA: Milford House, 1972), 280. Jonathan Simon, *Chemistry, Pharmacy, and Revolution in France, 1777–1809* (Burlington, VT: Ashgate Publishing Company, 2005), 40.
4 Glenn Sonnedecker, *Kremers and Urdang's History of Pharmacy*, 43.
5 A.C. Wootton, *Chronicles of Pharmacy, Vol. 1*, 278. Charles Lawall, *The Curious Lore of Drugs and Medicines: Four Thousand Years of Pharmacy* (Garden City, NY: Garden City Publishing, 1927), 364–365.
6 Ibid., (Wootton), 277–278.
7 Charles Lawall, *The Curious Lore of Drugs and Medicines*, 371–372. Glenn Sonnedecker, *Kremers and Urdang's History of Pharmacy*, 355.
8 Ibid., (Lawall), 372–373.
9 Glenn Sonnedecker, *Kremers and Urdang's History of Pharmacy*, 355.
10 Ibid., 355. Charles Lawall, *The Curious Lore of Drugs and Medicines*, 373–374.
11 Ibid., 376–378. A.C. Wootton, *Chronicles of Pharmacy, Vol. 1*, 266–270. George Bender, *Great Moments in Pharmacy* (Detroit, MI: Northwood Institute Press, 1966), 88–91.
12 Ibid. (Wootton), 271–272. Glenn Sonnedecker, *Kremers and Urdang's History of Pharmacy*, 356.
13 Rudolf Schmitz, "Friedrich Wilhelm Serturner and the Discovery of Morphine," *Pharmacy in History*, Vol. 27 (1985), No. 2, 61–74.
14 Ibid., 62. George Bender, *Great Moments in Pharmacy*, 99. A.C. Wootton, *Chronicles of Pharmacy, Vol. 2* (Reprint, Boston, MA: Milford House, 1972), 244–245. James Grier, *A History of Pharmacy* (London: The Pharmaceutical Press, 1937), 93.
15 Jenny Sutcliffe and Nancy Duin, *A History of Medicine* (New York: Barnes and Noble Books, 1992), 50–51.
16 Diarmuid Jeffreys, *Aspirin: The Remarkable Story of a Wonder Drug* (New York: Bloomsbury Publishers, 2004), 38. A.C. Wootton, *Chronicles of Pharmacy, Vol. 2*, 248–249. George Bender, *Great Moments in Pharmacy*, 100–103.
17 David L. Cowen and William H. Helfand, *Pharmacy: An Illustrated History* (New York: Harry Abrams, 1990), 125.

9 Colonial and Early American Pharmacy

How did the apothecaries who lived during these eras shape the practice of American pharmacy?

One of the main motivations for the Age of Exploration was for the nation-states of Western Europe to find new trade routes to the Far East. By doing so, the Western Europeans found routes that were able to break the Italian city-states' long-standing monopoly on products (spices, herbs) coming from Asia, Africa, and the Middle East. When Christopher Columbus and other European explorers searched for these new trade routes they stumbled upon the Americas, and for them, had discovered a New World laden with new resources, especially *materia medica*. So, the "discovery" of the Americas became an unintended consequence of the Western Europeans' desire for better access to products from the East and represented a vast new territory ready for them to exploit commercially, which they did. The bounty of flora and fauna found in the New World offered the Europeans a seemingly endless supply of new materials that would dramatically expand *materia medica* in Europe and elsewhere in the Old World. Potatoes, tomatoes, corn, cotton, turpentine, peanuts, tobacco, ipecacuanha, cinchona, sarsaparilla, and other plants were shipped back to Europe.[1]

The New World held great promise for its settlers, but also held great risks and danger. As Paul Starr has noted in his landmark book, *The Social Transformation of American Medicine*, the colonists faced isolation and geographic vastness on an unprecedented scale. The colonies started with very sparse populations that could not support specialized medical people even if they had the means to attract them. So, self-reliant, do-it-yourself medicine would have to suffice and this pragmatic approach remains a persistent theme in American medicine to the present day. By necessity and mercantilist policies, colonists imported books and medicines from Europe whenever they could, but over time learned about the health benefits of the bountiful flora and fauna that surrounded them.[2] Until hostilities took hold, the colonists learned about Native American medicine and used it with great benefit to supplement traditional European medicine. In fact, there are still

170 botanical drugs officially listed in the *United States Pharmacopoeia* or the *National Formulary* that Native Americans who lived north of the Rio Grande river used. There are another 50 botanical drugs that remain officially listed that derive from Native Americans who lived south of the Rio Grande river.[3] Because of the lack of trained medical people, practitioners often played many roles, shifting from physician to surgeon to apothecary in order to serve the variety of health needs of their patients. In 1776, during the American Revolution era there were about 3,500 practicing physicians with about 500 of them holding European degrees.[4] Due to all of the reasons listed above, the classic Kremers and Urdang textbook aptly refers to this era as "The Period of Unorganized Development."[5]

Pharmacy in New France

With sea voyages lasting for months under precarious conditions, only the most adventurous people were willing to sail to the New World. One such intrepid character was Louis Hebert (*c*.1575–1627), the son of the apothecary to the French royal family. As a boy Hebert had been thrilled by the stories of Jacques Cartier's voyages up the St. Lawrence River. In 1604, while working in his apothecary shop in Paris, Hebert was offered a place on an expedition to settle New France by Simon de Monts who led a fur-trading company. One of Hebert's fellow voyagers was the French explorer Samuel de Champlain (1567–1635) who believed that it was essential for the French to send settlers to New France to colonize it quickly. The 50 settlers landed in Western Nova Scotia across from the modern city of Annapolis Royal and called their settlement "Habitation." Habitation lasted from 1606 until 1613, when it was destroyed by the English. While at Habitation, Hebert took care of the settlers' medical needs and learned about the native plants from the Micmac Indians. Some of the plants he learned about were Boneset, Mullein, Jack-in-the Pulpit, and Golden Seal. After the destruction of the settlement, Hebert returned to Paris and reopened his apothecary shop, but longed for the adventure that the New France offered. In 1617, Champlain asked Hebert if he was interested in settling in Quebec. Hebert took his family and spent the rest of his life tending to the medical needs of his fellow colonists, learning about native medicinal plants, and farming. In 1627, Hebert died as a result of an accident. Today, there is a statue near city hall in Quebec that honors perhaps the first pharmacist in North America for his pioneering work and spirit.[6]

Pharmacy in New Sweden and New Netherlands

To the south of New France, Henry Hudson sailing for the Dutch reached the New World at a harbor the Dutch would call New Amsterdam (later New York). Searching for the elusive "Northwest Passage" Hudson sailed up the river that was later named in his honor, giving the Dutch a claim to North American territory they called New Netherlands. The Swedes established a colony south of New Amsterdam (modern day New Jersey and Delaware) which they called New Sweden. It was settled more by Finns than Swedes. The Scottish apothecary, John Johnstone, founded a settlement in what later became Perth Amboy, New Jersey in 1638. In 1655, the Dutch conquered New Sweden and annexed it. The first surgeon of note in New Amsterdam was Gysbert van Imbroch (?–1665),who operated out of a shop in 1653 where he practiced medicine and sold drugs as part of a general store. In 1663, he moved up to the Hudson River to a town called Wildwyck (renamed Kingston in 1669 when the English conquered New Netherlands), which he operated until his death in 1665. While it might have been a barber-surgeon's shop, it also sold drugs and may have been one of the first drugstores in North America. Imbroch's "drugstore" was more of a dispensary that served as a place for him to prepare prescriptions and sell them to his patients as part of his surgical practice. These types of multi-purpose dispensaries or "doctor's shops" would become the early models of pharmacy practice in the North American colonies.[7]

Pharmacy in the British North American Colonies

Not to be outdone by the French or Dutch in the race for colonies, the English established colonies in North America first at Jamestown, Virginia, and then in Massachusetts. In 1607, Jamestown was founded and in 1608 there were two apothecaries who settled here, Thomas Field and John Harford, but the records make no further mention of them. There was a record of a Dr. Pott in the Virginia colony who was sued by his apothecary apprentice, Richard Townsend, who complained that he did not receive the training he had been promised. The court ordered Pott to fulfill his promise or compensate Townsend.[8] The scant records indicated that the physicians and surgeons often prepared their own prescriptions with a few willing to take on apothecary apprentices.

The Pilgrims and Puritans in New England

In 1620, the Pilgrims landed and established Plymouth colony with a Dr. Samuel Fuller aboard.[9] They were followed by the Puritans who landed in Salem in 1628. In 1630, several hundred more settlers arrived and settled in Boston, which was founded by their governor, John Winthrop (1587–1649). As governor, it was certainly in Winthrop's interest to promote health among the colonists to insure the survival of the colony. Winthrop and his son John Winthrop, Jr. (1606–1676), who later became governor of Connecticut Colony, took an avid interest in preparing medicines. The Winthrops were assisted in preparing medicines by a trained apothecary from England named Robert Cooke (*c*.1615–1640s) who arrived in 1638 and worked into the 1640s. In 1643, they received excerpts and interpretations taken from John Gerard's *Herball* from an Edward Stafford of London. The younger Winthrop went beyond importing herbs from Europe and took an interest in chemistry, preparing compounds from saltpeter, antimony, mercury, tartar, sulfur, and iron. He also became skilled in preparing galenicals and became one of the first people in North America to engage in preparing pharmaceuticals.[10] So it was no accident that the first apothecary shops or dispensaries opened in Massachusetts.

The first woman to practice as an apothecary during the colonial era was Elizabeth Gooking Greenleaf (1681–1762). She was born in Boston to a prominent colonial family, and in 1699 married Daniel Greenleaf and had 12 children with him. Greenleaf, a Harvard educated minister, physician, and apothecary became the pastor of the Congregational Church in Yarmouth, Massachusetts. Elizabeth helped her husband prepare medicines for his patients and in 1727 they opened an apothecary shop in Boston where she had the distinction of being the only woman apothecary among the 32 apothecaries serving New England during the late seventeenth and early eighteenth centuries.[11]

The first mention of a pharmacy in British North America was the dispensary of William Davis (Davice) (*c*.1617–1676) of Boston in 1646.[12] The first records of the daily operations of a dispensary came from the account book of Bartholomew Browne whose shop was located in Salem in 1698. Apparently, Browne derived most of his income from selling prescriptions, but interestingly also charged patients for medical advice as had been the practice for British apothecaries.[13] Dr. Zabdiel Boylston, who gained fame for inoculating patients to prevent smallpox almost 70 years before Edward Jenner did, owned and operated a highly successful apothecary's shop in Boston. As early as 1723 Boylston advertised some of his herbs in *The Boston Gazette* including, among others, saffron, jalap, cassia, and juniper berries.[14] By 1721, there were 14 apothecary shops that were owned by physicians (doctor's shops) that in addition to drugs, sold other commodities including paints, varnishes, glass, and wallpaper, becoming general stores.[15]

Pharmacy in Philadelphia

Christopher Marshall (1709–1797), an Irish immigrant from Dublin, founded an apothecary shop in Philadelphia that would shape the course of American pharmacy for 96 years (1729–1825). Beginning in 1729, Marshall's shop expanded in 1735 as a manufacturer of pharmaceuticals and, over time, would become an on-the-job training center for aspiring apothecaries who wanted to be on the cutting edge of pharmacy. Marshall became a wholesaler (druggist) of medicine chests to physicians and large landholders with a small sideline business retail. In 1765, Marshall made his sons Christopher Marshall, Jr. (1740–1806) and Charles Marshall (1744–1825) partners in the family business, instilling in them the importance of service to their community. Marshall and his store played a key role during the American Revolution, supplying medicines to George Washington's troops in Pennsylvania, New Jersey, Delaware, Maryland, and Virginia. In 1776, Marshall received a commission to tend to the needs of the wounded and sick in the hospitals of Philadelphia. Marshall's sons embarked on a large-scale manufacturing operation, preparing medicine chests to support the patriot side during the American Revolution.[16]

After Charles retired from the business, his son made some poor business decisions and the shop fell on hard times until Charles's daughter, Elizabeth, took over its management in 1805. Elizabeth Marshall, the first American woman apothecary, restored the family business, and it flourished until she sold it in 1825.[17] Perhaps the Marshall family's most significant contribution

Figure 9.1 Christopher Marshall (1709–1797), an Irish immigrant, opened the premier apothecary shop in Philadelphia in 1729, which would influence the course of American pharmacy.

Source: Painting by Robert Thom, reproduced courtesy of the American Pharmacists Association Foundation.

to pharmacy occurred when Charles Marshall went on to become the first president of the Philadelphia College of Pharmacy, serving from 1821 until his death in 1825. The values of meticulous pharmaceutical skill, irreproachable integrity, and dedication to patient well-being became a lasting legacy in American pharmacy, in part, due to the efforts of the Marshall family.[18]

Pennsylvania Hospital, John Morgan, and the Birth of American Pharmacy

The practice of pharmacy as a separate and distinct field of medicine got its first boost with the founding of North America's first hospital in Philadelphia. Dr. Thomas Bond and Benjamin Franklin (1706–1790) established Pennsylvania Hospital in 1751 in a rented house. The original hospital consisted of eight beds with four patients. Due to an over-shipment of drugs from London, the drugs were stored in a specially designated room and thus the first American hospital pharmacy was established. In 1756, the first permanent part of the hospital was built on its current site and forms the east wing of today's hospital. With the hospital pharmacy in place, Dr. Jonathan Roberts was hired

Figure 9.2 John Morgan (1735–1789), a life-long Philadelphian, he pioneered medical education in the United States and promoted the practice of physicians prescribing medicines and apothecaries preparing them.

Source: Painting by Robert Thom, reproduced courtesy of the American Pharmacists Association Foundation.

in 1752 at a salary of 15 pounds per year to serve as its apothecary. Roberts served for a year and was replaced by Dr. John Morgan (1735–1789), where he too served for a year, before embarking on a career in medicine.[19]

After graduating from the College of Philadelphia (now the University of Pennsylvania) in 1757 he went on to Europe in 1760 to study medicine. He graduated from Edinburgh in 1765 and published *Discourse Upon the Institution of Medical Schools in America* in which he called for the organization of rigorous medical education in the Americas, which led to the founding of the Medical Department of the University of Pennsylvania. In this work, Morgan called for a hierarchy and division of labor among physicians, surgeons, and apothecaries, arguing that each specialty required unique talents and training. He also argued that pharmacy required a special dedication to the art of compounding and to patients that was much more than a business. He called upon his fellow physicians to write prescriptions and follow the European model of medical and pharmacy practice. Although these views would form the basis of medical and pharmacy practice later, his views were met with great skepticism and caused significant controversy during his time. Upon his return to the Philadelphia, Morgan hired a Scottish apothecary to prepare his prescriptions, but his patients balked at the additional time and expense this model engendered and he was forced to revert to the former doctor's shop model.[20] Morgan taught several subjects, his favorites being the preparation of medicines and chemistry. Due to his distinguished service as a lieutenant during the French and Indian War in 1775, the Continental Congress appointed Morgan as "Director General and Physician in Chief" of the American Army. He called for rigorous examinations for army medical officers and appointed surgeons as hospital chiefs, which they resented. Complaints and political intrigues led Morgan to be relieved and although he was later vindicated by Congress, he went back into private practice and died a broken man in 1789, misunderstood by many of his contemporaries for his vision.[21]

Andrew Craigie: Apothecary General

One of John Morgan's wisest decisions was to appoint Andrew Craigie (1754–1819), a Boston apothecary, as America's first Apothecary General in 1777. Not much is known about Craigie's childhood other than the fact that he was educated at the Boston Latin School. Although little is known about his later education or apprenticeship, Craigie must have had some medical standing, as he was appointed commissary of medical stores by the Massachusetts Committee of Safety on April 30, 1775. He was tasked to provide "bedding and other necessities for the sick."[22] Craigie's resourcefulness and skill were put to the test during the Battle of Bunker Hill (actually Breed's Hill) on June 17, 1775, as he assisted Dr. D. Townsend in treating the sick and wounded at a makeshift medical station just behind the American lines. Due to the lack of medicines and supplies, Craigie had to call upon private citizens to share what they could to support the American soldiers.[23]

As tensions mounted between the American colonists and the British, on July 17, 1775 the Massachusetts Provincial Congress appointed Craigie as the medical commissary and apothecary to the Massachusetts militia that was being raised. This appointment marked the first time an apothecary became an officially recognized part of an American army. As the American Revolution began in earnest in 1775, the Continental Congress in Philadelphia began organizing a resistance and the organization plan created positions for apothecaries to serve in each army hospital. In 1777, as part of a reorganization plan, the Continental Congress created the position of Apothecary General at the rank of Lieutenant Colonel for four district departments in the colonies.[24]

Craigie was appointed Apothecary General to the northern district, receiving five pounds per month as pay. His primary responsibilities included procuring, preparing, and distributing medicine to the troops in his district. In 1778, Craigie established an Elaboratory in Carlisle, Pennsylvania that could produce, store, and distribute the medicines the Continental army needed to keep fighting. This appointment marked a first in the history of American pharmacy, because it recognized an apothecary as a commissioned officer, and the apothecary's job description was clearly articulated for the first time as distinct and separate from that of the physician. Both of these firsts set important precedents for American pharmacists in their quest to gain recognition as professionals throughout the nineteenth and twentieth centuries in the United States.[25]

In July 1780, the Continental Congress reorganized military medicine, replacing the four Apothecaries General with one apothecary and five assistants. Craigie received this appointment as chief apothecary and served until he left the service in November 1783. After the war, Craigie parlayed his connections with luminaries such as Alexander Hamilton and Henry Knox to become the Director of the Bank of the United States. Craigie became a wholesale druggist in New York, but the business failed in 1789 and he turned to real estate speculation. Involved in a number of speculative real estate deals, Craigie returned to Boston in 1791. He married Elizabeth Shaw in 1793, but the marriage failed. He lived as a recluse in the Vassal Mansion he had purchased in 1792 and died in relative isolation, hounded by debt collectors, in 1819.[26]

Hugh Mercer: American Revolution Apothecary and Hero

Another important figure who served with distinction during the American Revolution was the physician/apothecary Hugh Mercer (1725–1777). Born in Scotland in 1725, Mercer settled in Greencastle, Pennsylvania (now Mercersburg) and began his medical practice where he prepared his own medicines for his patients. Mercer served in the British army during the French and Indian War and was wounded at the battle of Fort Duquesne, the same battle in which General Braddock had been killed. Mercer met and befriended another aspiring colonial officer, George Washington. After the war, Mercer established a practice in Fredericksburg, Virginia in 1764 and opened an apothecary shop with a partner whose name is only known to history as Clement. Mercer's shop thrived and became the center of town, especially popular with visitors. Mercer's old war friend, George Washington, had a desk in the shop where he managed his land interests when in town from 1764–1776.[27]

When the American Revolution began, Mercer was commissioned as a Colonel in the Continental army. He rose to the rank of Brigadier-General and, as part of Washington's army, during the Battle of Princeton on January 3, 1777, led his detachment of riflemen against two British regiments before they could reach Washington's main force near Trenton. Mercer's riflemen could not withstand the British bayonet counterattack and fell back. Mercer's horse had been killed and the general was on foot slashing with his sword when he was struck down by a musket butt and knocked to the ground. The British thought Mercer was Washington and called for him to surrender, but Mercer responded with his sword at which point he was stabbed several times by bayonets. At this point, Washington rode onto the scene and rallied his troops on to a victory over the retreating British, chasing them out of New Jersey.[28] Mercer died from his wounds nine days later, on January 12, 1777. The Continental Congress authorized the erection of a monument to honor him that stands today in Fredericksburg.[30]

Just as warfare brings out the best in people, as evidenced by the case of Hugh Mercer, it also has a penchant for bringing out the worst, as evidenced by the case of Benedict Arnold. In his youth in Connecticut, Arnold learned about the apothecary's art in Norwich and later owned and operated a drugstore in New Haven from 1764, one of his many business ventures.[29] Arnold, of course, was an American general who had been in charge of West Point when he betrayed the American cause by selling secrets to the British and joined the British army.

Pharmacy During the American Revolution

Revolutions by definition entail rapid transformations that impact nearly every aspect of a society, and the American Revolution proved to be no exception. Being a war of independence, the American colonists fought the world's most dominant empire, the British, and won. The war for independence became an unprecedented experience for the colonists that forced the

Continental Congress to experiment in creating institutions that could withstand and overcome the British attacks. As we have seen, the Continental Congress created the offices of Director General and Chief Physician and Apothecary General in order to serve the medical needs of its new army, and it was compelled by the force of war to improvise rapidly to make these institutions work. The creation of this American military medical establishment had a profound effect in shaping the course of American pharmacy. For example, in 1778 Andrew Craigie, the Apothecary General for the northern district, envisioned and supervised the construction of the Elaboratory in Carlisle, Pennsylvania, to engage in large-scale manufacturing of drugs to fill the medicine chests for distribution to army hospitals and field units. Although herbal supplies were scarce, Craigie managed to make the Elaboratory work, sometimes substituting scarce European herbs with local ones. The typical American Revolution era medicine chest included calomel, Epsom salts, Peruvian bark (cinchona), and tartar emetic, as well as a mortar and pestle and a pewter syringe.[31] Depending on their availability, the medicine chests might contain opium or paregoric elixir, jalap, rhubarb, and Glauber's Salts.[32] Although the Congressional Act of 1790 called for standardization of the contents of these medicine chests, in practice the contents often varied until 1876 when the Marine Hospital Service, which later became the U.S. Public Health Service, took on this task.

The American Revolution also led to the development of the first American Formulary, the *Lititz Pharmacopoeia* written in Latin by Dr. William Brown (1752–1792) in Lititz, Pennsylvania in 1778. The pharmacopoeia consisted of 100 formulas; 84 for internal use and 16 for external use. Dr. Brown had been educated at the University of Edinburgh and undoubtedly was familiar with the *Pharmacopoeia Edinburgensis* published in 1756 and the *Pharmacopoeia Londinesis* published in 1746. The *Lititz Pharmacopoeia* allowed for substitutions of ingredients, reflecting the realities of the scarcities of ingredients due to the exigencies of the war, which harks back to earlier times when ingredients were also scarce. This pharmacopoeia also included formulas that called for North American herbs such as sassafras which the European pharmacopoeias did not. The *Lititz Pharmacopoeia* also distinguishes those medicines that must be made in large-scale laboratories with specialized equipment as opposed to those that could be compounded in field hospitals or dispensaries, once again reflecting the austerity imposed by the war.[33]

When France officially became America's ally after the Battle of Saratoga, Jean-François Coste, the Chief Physician to the French expeditionary army introduced another official formulary for use in military hospitals. Coste's formulary was called the *Compendium Pharmaceuticum, Militaribus Gallorum Nosocomilis, In Orbe Novo Boreali Adscriptum*, which was printed in Newport in 1780. Coste's formulary was steeped in the French pharmaceutical tradition and was based on the *Codex Medicamentarius seu Pharmacopoeia Parisensis* which was issued in 1758.[34] In addition to the French pharmaceutical tradition influencing American pharmacy, the British had hired over 13,000 Hessian mercenary troops from Prussia. Similar to the French, the Hessians brought their pharmaceutical and medical traditions with them to American soil.[35] After the war, American pharmacy took a decidedly patriotic turn, although it did manage to incorporate many formulas from foreign sources into its own pharmaceutical tradition, culminating in the publication of the *United States Pharmacopoeia* in 1820.[36]

American Revolution Therapeutics and Medications

The American Revolution also ushered in a new era of medical treatment called "heroic medicine." Dr. Benjamin Rush (1745–1813), a signer of the "Declaration of Independence," who became the first noteworthy physician in the United States was a fierce advocate of heroic medicine. Rush viewed disease as a pernicious enemy of humankind and therefore called for aggressive treatment to eliminate it. For Rush, this meant the aggressive bleeding of patients combined with high doses of calomel and jalap (or other laxatives such as castor oil) to achieve a maximum purging of the bowels to rid the body of disease. Just as his mentor Dr. William Cullen of Scotland had taught him, Rush viewed all disease as originating from the singular cause of the excitement of the capillaries, which was best cured by bleeding. This treatment was often augmented by purging, vomiting, and blistering that required administering harsh drugs. Rush wrongly believed the body contained more blood than it actually did. For Rush, it was incumbent upon the responsible physician to take every possible measure to heal patients. When other physicians fled the great yellow fever epidemic of 1793 in Philadelphia, Rush remained determined to fight the epidemic with every heroic measure he could take to save his patients. Rush was a physician of many achievements and his most lasting legacy was his staunch belief in the benefits of heroic medicine, and it became a hallmark of American medical practice until the Civil War and the advent of germ theory.[37]

Rush was born on December 24, 1745 in Byberry, Pennsylvania, on the family farm. After his father died when Rush was six, his uncle, Reverend Sam Finley, looked after his education. Rush graduated from the College of New Jersey (now Princeton University) in 1760 and worked as an apprentice to a Philadelphia physician, Dr. John Redman. This apprenticeship exposed Rush to the intellectual life at the Pennsylvania Hospital where he attended lectures by Dr. John Morgan and Dr. William Shippen, Jr. Taking his mentor's advice, Rush attended the University of Edinburgh where he studied under Dr. William Cullen and graduated in 1768. In 1769, Rush returned to Philadelphia to practice and received an appointment to teach chemistry at the College of Philadelphia (now the University of Pennsylvania). A demanding instructor, Rush wrote an influential textbook entitled, *Syllabus of a Course of Lectures on Chemistry*, in 1770. According to George Bender, during the 43 years Rush taught (1769–1812) he influenced over 3,000 aspiring physicians and apothecaries with his ideas about the efficacy of heroic medicine.[38] Rush died on April 13, 1813 as America's most prominent physician, a signer of the "Declaration of Independence," and Treasurer of the United States Mint.

American Revolution Era Drugs

Flowing from heroic medicine, a American Revolution era apothecary or physician had a number of medications that could be used to treat the symptoms of various ailments. Indeed, eighteenth- and early nineteenth-century apothecaries and physicians thought in terms of drugs that could be used to treat symptoms, and consequently the drugs were classified that same way. Anodynes were used as pain relievers and included opium and laudanum (which was composed of opium, saffron, and Canary wine). Anti-arthritics were used to treat pain with inflammation and included Epsom salt or cinchona (quinine for fever or ague). Anti-dysentery medicines included ipecac mixed with blackberry wine, and paregoric or elixir asthmaticum, which was composed of opium, honey, licorice, benzoic acid, camphor, oil of anise, potassium carbonate, and alcohol. Anti-pyretics or febrifuges were treatments to prevent or reduce fever and included the use of emetics, cinchona, laxatives, and cold baths. Emetics or vomiting agents were used to treat various forms of food poisoning and included tartar emetic, ipecac, warm water, and honey. Muscle spasms were treated with opium, wine, cinchona, and oil of amber. For intestinal irritation purgatives or cathartics were used that included Glauber's Salts, Plummer's Pills, ipecac, jalap, calomel, salme, rhubarb, castor oil, and Epsom salts. Sometimes mercury in oil was used to increase salivation to counter intestinal irritation. Sudorifics or diaphoretics were used to induce perspiration and included camphor, Dover's Powder (opium and ipecac), and rhubarb. Diuretics to treat edema (dropsy) were used to increase urine flow and included milk, extracts of dandelions, juniper berries, and lemon juice.[39] From a modern standpoint, the treatment of illness during this era was often worse than the illness itself.

Chapter Summary

Only the boldest or most persecuted people ventured forth from Europe into the New World. Confronted by a seemingly limitless land of immense resources the colonists operated with a pioneering spirit that they would make a life for themselves in this new land, no matter what it took. This spirit of independence manifested itself and became accentuated by the British colonists in the American Revolution. During the war for independence, new institutions such as the Continental Congress would define a new medical establishment for the United States that was based upon European models, but would be operated in a uniquely American way. For pharmacy, it meant an opportunity to be recognized as a separate and important part of medical establishment. The *Lititz Pharmacopoeia* borrowed from its European predecessors, but proudly included indigenous botanicals in its formulary. The Marshall family and Andrew Craigie's Elaboratory showed that large-scale manufacturing of pharmaceuticals had a place in the United States, as old as the new Republic itself. With the success of the American Revolution, American pharmacy had been born.

Key Terms

Louis Hebert	medicine chests	*Lititz Pharmacopoeia*
Gysbert van Imbroch	John Morgan	Coste's *Compendium*
"doctor's shops"	Andrew Craigie	Benjamin Rush
John Winthrop, Sr.	Elaboratory	"heroic medicine"
Christopher Marshall	Hugh Mercer	calomel
Elizabeth Marshall	Benedict Arnold	jalap

Chapter in Review

1   Describe the impact of the discovery of the New World on the development of pharmacy.
2   Explain the conditions the colonists faced.
3   Describe the unique aspects of the development of pharmacy in British colonial North America.
4   Compare and contrast the highlights of the development of pharmacy in Boston and Philadelphia.
5   Explain the career of Dr. John Morgan as a pioneer of American pharmacy.
6   Trace the development of the birth of American pharmacy as a result of the American Revolution.
7   Explain the important impact Dr. Benjamin Rush had on the rise of heroic medicine in America.

Notes

1 George E. Osborne, "Pharmacy in British Colonial America." In *American Pharmacy in the Colonial and Revolutionary Periods*, edited by George A. Bender and John Parascandola (Madison, WI: American Institute of the History of Pharmacy, 1977), 5–6.

2 Paul Starr, *The Social Transformation of American Medicine* (Princeton, NJ: Princeton University Press, 1982), passim.

3 Virgil J.Vogel, *American Indian Medicine* (Norman: University of Oklahoma, 1970), 318. David Armstrong and Elizabeth Metzger Armstrong, *The Great American Medicine Show* (New York: Prentice-Hall, 1991), 17–18.

4 Charles Lawall, *The Curious Lore of Drugs and Medicines: Four Thousand Years of Pharmacy* (Garden City, NY: Garden City Publishing, 1927), 107.

5 Glenn Sonnedecker, comp., *Kremers and Urdang's History of Pharmacy* (4th ed., Madison, WI: AIHP, 1976), 144.

6 Ibid., 149–150. George Bender, *Great Moments in Pharmacy* (Detroit, MI: Northwood Institute Press, 1966), 72–75.

7 Glenn Sonnedecker, *Kremers and Urdang's History of Pharmacy*, 151–152.

8 George E. Osborne, "Pharmacy in British Colonial America," 7. George Urdang, "Pharmacy in the United States in Colonial North America" (*The Merck Report*, April 1947), (Reprint: Madison: AIHP, 1976), 3.

9 Ibid. (Osborne), 8.

10 George A. Bender, *Great Moments in Pharmacy*, 76–79. Robert C. Winthrop, "Receipts to Cure Various Disorders: For my worthy friend Mr. Winthrop. 1643." In *The Badger Pharmacist*, (Madison, WI: Wisconsin Chapter of Rho Chi, No. 15, April 1937), 1–24.

11 Metta Lou Henderson, *American Women Pharmacists: Contributions to the Profession* (New York: Pharmaceutical Products Press, 2002), 2.

12 Norman Gevitz, "Pray Let the Medicines Be Good: The New England Apothecary in the Seventeenth and early Eighteenth Centuries." In *Pharmacy in History*, edited by Gregory J. Higby (Madison, WI: AIHP),Vol. 41 (1999), No. 3, 90.

13 Ibid., 87.

14 George Griffenhagen, *Bartholomew Browne, Pharmaceutical Chemist of Salem 1698–1704* (Essex: Essex Institute Historical Collections, Jan. 1961), 19–30.

15 George E. Osborne. "Pharmacy in British Colonial America," 10.

16 George A. Bender, *Great Moments in Pharmacy*, 80–83.

17 Metta Lou Henderson, *American Women Pharmacists*, 3.

18 Ibid., 83.

19 Glenn Sonnedecker, *Kremers and Urdang's History of Pharmacy*, 160–161.

20 Gregory J. Higby, "From Compounding to Caring: An Abridged History of American Pharmacy." In *Pharmaceutical Care*, edited by Calvin H. Knowlton and Richard P. Penna (2nd ed., Bethesda, MD: ASHP, 2003), 21–22.

21 Glenn Sonnedecker, *Kremers and Urdang's History of Pharmacy*, 164.

22 George Bender, *Great Moments in Pharmacy*, 92.

23 Ibid., 94.

24 Ibid., 94.

25 Dennis Worthen, *Heroes of Pharmacy: Professional Leadership in Times of Change* (Washington, DC: American Pharmacists' Association, 2008), 60–63.

26 Ibid., 61–62.

27 George Urdang, "Pharmacy in Colonial North America," 7.

28 C. Keith Wilbur, *Revolutionary Medicine 1700–1800* (Chester, CT: The Globe Pequot Press, 1980), 67–68.

29 George Urdang. "Pharmacy in Colonial North America," 7.

30 Charles Lawall, *The Curious Lore of Drugs and Medicine*, 409.

31 Dennis Worthen, *Heroes of Pharmacy* (Washington, DC: APhA, 2008), 60.
32 George Griffenhagen, "Drug Supplies in the American Revolution." In *United States Museum Bulletin 225* (Washington, DC: Smithsonian Institution, 1961. Reprint, Madison: AIHP, 1976), 130–133.
33 Ibid., 110. Glenn Sonnedecker, *Kremers and Urdang's History of Pharmacy*, 170. George Urdang, "Pharmacy in the United States Prior to the Civil War." In *The Merck Report*, April 1947 (Reprint, Madison: AIHP, 1976), 8.
34 Ibid., 8.
35 Rudolf Schmitz, "The Medical and Pharmaceutical Care of Hessian Troops During the American War of Independence." In *American Pharmacy in the Colonial and Revolutionary Periods*, edited by George Bender and John Parascandola (Madison, WI: AIHP, 1976), 40.
36 Lee Anderson and Gregory J. Higby, *The Spirit of Voluntarism: A Legacy of Commitment and Contribution, The United States Pharmacopoeia 1820–1995* (Rockville, MD: United States Pharmacopoeial Convention, Inc., 1995). Glenn Sonnedecker, "The Founding of the Pharmacopoeia." In *Pharmacy in History* (Madison, WI: AIHP), Vol. 35 (1993), No. 4, 149–200.
37 George Bender, *Great Moments in Medicine* (Detroit, MI: Northwood Institute Press, 1965), 152–153.
38 Ibid., 157.
39 C. Keith Wilbur, *Revolutionary Medicine, 1700–1800*, 12.

10 The Era of Alternative and Patent Medicine

How did alternative (complementary) medicine originate and why does it continue to appeal to many Americans today?

In the wake of the American Revolution there was a populist backlash against mystical and paternalistic European medicine and pharmacy that smacked of colonialism. In particular, there was a populist backlash against heroic medical treatments that called for bleeding and purging coupled with harsh chemicals and herbs. This ushered in the era of self-help and do-it-yourself, common sense medicine that developed a wide following, especially in frontier areas. Congress had passed patent legislation to promote the spirit of innovation and private enterprise in a free market guided by the principle of laissez-faire. The rise of newspaper advertising also promoted, and to some extent legitimized and increased, the sales of patent medicines, despite the warnings from the medical community. Drug advertising dates back to the days of the early republic and remains a mainstay of American life today. Only the drugs and the types of media have changed—advertising has persisted as the lifeblood of American media. The Age of Advertising was born during this era and played a key role in shaping the course of American pharmacy.

The Birth of Alternative Medicine in America

The development of alternative medicine in the early republic was shaped by several influences of that time. Of course, the frontier mentality viewed nature as something to be conquered and disease was no exception. Living on the frontier meant living a life of necessity when home health care was the only kind available. The Scottish physician, Dr. William Buchan (1729–1805), wrote an immensely popular book, *Domestic Medicine*, which was published in 1769 in Edinburgh and then was published in Philadelphia in 1771. The book was written in plain language and urged its readers to monitor their diets, practice hygienic habits, and exercise temperance. The book also contained a number of simple remedies that most households could use for a variety of ailments. Buchan saw his book as, "an attempt to render the Medical Art more generally useful, by showing people what is in their own power both with respect to the Prevention and Cure of Diseases."[1] He argued that, "No discovery can ever be of general utility while the practice of it is kept in the hands of a few."[2] This egalitarian attitude found an eager and grateful audience in the newly created republic. After 30 editions following its first publication in the United States, Buchan's book was replaced by a similar book that updated the original material but kept the same title. John C. Gunn's *Domestic Medicine* appeared in 1830 and sold quite well for many years. Similar to Buchan's philosophy, Gunn maintained his book was written, "In Plain Language, Free from Doctor's Terms . . . Intended Expressly for the Benefit of Families . . . Arranged on a New Simple Plan, By Which the Practice of Medicine is Reduced to Principles of Common Sense."[3]

Another powerful current of thought that influenced medicine and pharmacy during this time was Romanticism. Early nineteenth-century Romanticism was a reaction to the Enlightenment's central tenet that human beings were essentially rational creatures who were capable of unlocking the mysteries of the universe using the scientific method. By contrast, the Romantics argued that human beings were essentially creatures of emotion whose inner lives and dreams guided their actions. The Romantics were skeptical of the Enlightenment's emphasis on reason and saw the limits of science and objectivity. Having seen the carnage and turmoil caused by the Radical Stage of the French Revolution supposedly committed in the name of reason, the Romantics searched for other answers. Thus, the Romantics were apt to look to the past to find answers to explain the human condition. Romantic ideals shunned large industrial cities in favor of nostalgic rural settings, which fed directly back into the American frontier mentality, especially the idea of Manifest Destiny. Buchan's and Gunn's books urging self-reliance embodied the bold individualistic spirit of the pioneers' subduing of the North American frontier.

When President Andrew Jackson (1767–1845) took office in 1829, his election marked a major shift in American political thought. Jackson was the first American president who was born on the frontier of the Carolinas and reflected the ideas and aspirations of the American frontiersmen. Jackson despised the Eastern establishment and was determined to rid the United States of any institutions that smacked of elitism and exclusive membership. This spirit of Jacksonian democracy inspired many experiments that would foster the development of alternative medicine in America. At least one group of medical sectarians, known as the Thomsonians, made claims that their movement and medicinal products had the personal support of President Jackson himself!

John Wesley's Primitive Physick

Another influence on the rise of alternative medicine was the Great Awakening that launched an era of American religious revivalism. John Wesley,

(1703–1791) the founder of British Methodism, stressed the Protestant ideal of the self-directed and autonomous individual in all things including medicine. In 1747, Wesley published *Primitive Physick* which was an inventory of ancient cures for diseases and injuries. For example, licorice was recommended to suppress coughs and toasted cheese was used to close wounds. Unlike Buchan's and Gunn's later books, Wesley only offered a list of treatments, with little description of diseases and their symptoms. Still, just as Buchan and Gunn later, Wesley denounced physicians and the medical community as obfuscators and hoarders of knowledge that they hid from the public, only parceling out bits and pieces to enrich themselves.[4]

The Thomsonians

Amid this post-revolution movement toward democratizing medical knowledge a number of individuals appeared, launching their own alternative systems of medicine. Samuel A. Thomson (1769–1843) was born on a farm in Alstead, New Hampshire, to devout Baptist parents. He was born with a club foot, but was sent out to work on the farm with his siblings. With no schooling other than working on the farm Thomson stumbled upon *Lobelia inflata*, a plant that grew wild in the fields. Curious, he chewed on it causing him to vomit. He had found the plant that became the hallmark of his system of medicine that would earn him fame, fortune, and great controversy later in life.[5]

Thomson became acquainted with local folk medicine from a widow named Benton who lived near the Thomsons. When he contracted a severe rash, Benton successfully treated it, which piqued Thomson's interest in the use of botanicals as healing agents. Benton nurtured Thomson's interest in folk remedies, taking him on outings to pick herbs and then showed him how to prepare them as medicines. During this time Thomson's mother became gravely ill with consumption (pulmonary tuberculosis) and underwent the traditional heroic medical treatments, including bleeding, blistering, and purging by local physicians. He witnessed his mother's suffering at the hands of the regular physicians and, when she died within a few weeks, Thomson developed a lifelong dislike of physicians and heroic medicine, yet he once tried to become an apprentice to a local root doctor, who rejected him due to his lack of formal education.[6]

In another encounter with heroic medicine, Thomson asked a local physician to treat his two-year-old daughter for a canker rash (scarlet fever), which was unsuccessful. The desperate father placed his daughter over steaming water with a damp cloth to shield her, and she recovered from the rash fully, although she became blind in one eye.[7] Thomson concluded that he could cure people of illnesses while doing much less harm than the traditional physicians he came to regard as social parasites, and in 1824 he wrote a manifesto to this effect entitled, "Learned Quackery Exposed."[8]

SAM^{L.} THOMSON_ *BOTANIST.*

His System and practice, originating with himself.

Born Feby 9th 1769.

Figure 10.1 Samuel Thomson (1769–1843), a self-educated New England farmer who pioneered alternative medicine in the United States and founded a patent medicine business that influenced American commercial pharmacy for nearly a century.

Source: Courtesy of the National Library of Medicine.

In 1800, Thomson began practicing his new system of medical treatment, securing a federal patent on it on March 3, 1813 and a second in 1823. Thomson's theory was a simplified variation of Galen's humoral theory. According to Thomson, people contracted illness because of cold conditions. Thus, he believed that heat applied internally and/or externally would effect a cure.

Convinced he had discovered a breakthrough cure to human illness, Thomson visited Philadelphia to garner the endorsement of the chief proponent of heroic medicine, Dr. Benjamin Rush. Thomson met with Rush's associate Dr. Barton, but met only briefly with Rush, who according to Thomson seemed supportive of his ideas. Unfortunately, Rush died soon after their brief encounter.[9]

His first treatment option was to administer *Lobelia inflata* (Indian tobacco) laced with cayenne pepper that would "heat" the body up internally. Of course, the lobelia would cause the patient to vomit or, in Thomson's view, to purge themselves of poison in their system. He would then apply heat to the body by prescribing a steam bath for the patient to "heat" them up externally. While Thomson claimed in his patent that these treatments were new, Native Americans had been using them for millennia. As Thomson built his practice he employed about 70 herbs and drugs including camphor, ginseng, horse radish, peppermint, turpentine, and others.[10] A few of Thomson's remedies, tinctures of capsicum and myrrh, became accepted into the *United States Pharmacopoiea*.[11] Over time, Thomson's treatments became more elaborate, with the administration of enemas, infusions, powders, syrups, and tinctures. Thus, these treatments left patients vomiting, sipping hot teas, and/or sitting through long steam baths which proved to be marginally more pleasant for patients than the classic heroic treatment of bloodletting followed by large doses of calomel and/or jalap. Still, during this era the cure was often worse than the illness itself.

As Thomson's reputation as a healer grew, at least one local physician, a Dr. French, filed charges against him for killing one of his patients in 1809 by

Lobelia was named in honor of the Flemish botanist Matthias de Lobel (1538–1616) who had discovered the plant, also known as gag-root, pukeweed, or vomitwort.[12]

Thomson's main remedy *Lobelia inflata* was chemically analyzed by none other than William Procter, Jr. in his Philadelphia apothecary shop's lab where he isolated the active alkaloid in lobelia and named it "lobeline." Procter, who became known as the "father of American Pharmacy," published three scientific papers about the potential uses of lobelia and its alkaloids, but few outside of the botanical and Thomsonian communities ever pursued it.[13]

administering too much lobelia. Thomson was arrested and languished in a cold, lice infested prison cell until his trial. At his trial in Salem, Massachusetts, Thomson maintained his patient had left his bed and went outside, catching cold causing his demise. A prominent botanist, Manasseh Cutler (1742–1823), testified that the plant the prosecution presented as evidence was in fact marsh rosemary, and not *Lobelia inflata*. The case was dismissed and Thomson went back to his practice. Despite the outcome of the case, traditional physicians continued to lambast Thomson and his followers, referring to them in derogatory terms as "puke doctors" and "steamers." One pejorative verse against the Thomsonians went:

> I puke, I purge, I sweat 'em
> and if they die, I let 'em.[14]

In an age where there was little government regulation and entrepreneurship could grow unfettered Thomson's system of medicine developed great resonance among growing sectors of the American public. Thomson began selling family right certificates to his "Improved System of Botanic Practice" for the hefty price of 20 dollars. For a few dollars more franchisees could buy a copy of his autobiography, *A Narrative of the Life and Discoveries of Samuel Thomson*. During his life, Thomson sold nearly 100,000 of these family rights to his system. By this time, Thomson had moved to Boston and employed agents to market his system. Thomson struck a patriotic chord with the public by claiming his medicines were composed of American ingredients and not European imports. Thomsonian pharmacies competed directly with traditional pharmacies in preparing and dispensing Thomsonian medicines. Thomson's family rights system even allowed women franchisees to treat men, long before the first American woman physician Dr. Elizabeth Blackwell (1821–1910) began her practice in 1849 or before Mary Putnam graduated from the New York College of Pharmacy in 1863.[15]

In the early 1820s, Thomson published a number of books that became the cornerstone of the movement, including his self-help blockbuster *New Guide to Health* that was published in 1825. This volume was followed by his scathing attack on physicians, lawyers, and clergymen entitled, *Let Quackery be Exposed* (1824) in which he lambasted these professionals as social parasites who took advantage of the average person's suffering. Thomson's followers established a journal called the *Philadelphia Botanic Sentinel and Thomsonian Medical Revolutionist* that continued to attack traditional medicine while promoting their own cause. Friendly botanic societies emerged with the heart of Thomson's following coming from the Mid-West and rural South. The height of Thomson's movement came in the 1830s when they held annual conventions, the first of which was held in Columbus, Ohio in 1832 to celebrate the movement's success in treating cholera victims during the recent epidemic.

As Thomson's influence grew conflicts arose in the movement's ranks. Thomson learned the hard way that in a modern world of multiple perspectives schism often breeds schism and his movement proved to be no exception. With notoriety and large sums of money at stake, Thomson had a difficult time holding the movement that he inspired together. According to David and Elizabeth Metzger Armstrong, "By 1837, there were 167 authorized agents in 22 states and territories."[16] Some of Thomson's agents could not resist the temptation to start selling their own books and medicines under his name and pocketing the profits for themselves. Thomson's legal actions against his former minions did little to stop them from going into business for themselves amid a laissez-faire market. During the annual convention in 1838, one of Thomson's followers and editor of the *Thomsonian Recorder*, Alva Davis (1797–1880) led a faction that called for a more complete institutionalization of Thomson's system of medicine. This group called for more formal education and the establishment of Thomsonian medical schools. Thomson fulminated against such proposals and rejected them as outrageous. Consequently, Davis and his group walked out of the convention and started the Independent Thomsonian Botanic Society complete with a medical school in Columbus, Ohio. They later dropped Thomson's name and referred to themselves as Physio-Medicals. After Thomson's death in 1843, most of his followers gravitated toward a larger botanic medical sect known as the Eclectics led by Dr. Wooster Beach (1794–1868) from New York.[17] Thomson's system lived on in England due to the efforts of self-styled Thomsonian imitator named Albert Isaiah Coffin (1790–1866).[18] In the end, the contradictions in Thomson's system of medicine overwhelmed it. Although Thomson promoted his system as empowering his franchisees and patients, he held an exclusive patent on it and tried to maintain strict control of every aspect of it which, in the end, contributed to its demise.

The Eclectics

Unlike Samuel Thomson, Wooster Beach (1794–1868) was a formally educated physician who came to reject the practice of heroic medicine in favor of pragmatic herbal remedies and treatments that worked. Beach opened an infirmary in New York in 1827 and later a medical school in Worthington, Ohio, in 1830 that admitted African-Americans and women. Not as doctrinaire in his approach as the Thomsonians, Beach was willing to try any treatment that would work, from any source with the exception of the harsh minerals used in heroic medicine. In one of his books, the Philadelphia botanist Constantine Rafinesque (1783–1840) referred to Beach and his followers as "eclectics" because of their willingness to try anything to heal patients. Rafinesque also

saw great promise in the advances of botany and chemistry to produce medicines in a scientific manner and also saw a leading role for pharmacists to
play in this transformation. Beach published *The American Practice of Medicine* in
1833, which formed the basis of this medical and semi-political movement.[19]

Hoping to compete on a scientific level with traditional pharmacy and the
advent of alkaloid chemistry, the Eclectic physician and pharmacist John King
(1813–1893) developed concentrated drugs made with plant resinoids sus-
pended in an emetic known as May apple root. These drugs, which could be
mass produced, were easier to administer than traditional herbal compounds
and the dosages could be standardized. The Eclectics used over 350 medicines that were sold both in Eclectic and traditional pharmacies. Yet, like
Thomsonian medicines, these medicines were all emetics and by the 1860s
began to fall out of public favor. King teamed up with John Scudder (1829–
1894) to produce better tasting medicines with smaller doses. They hired one
of the most talented pharmacists in American history, John Uri Lloyd (1849–
1936), to help them. Lloyd was a nineteenth-century Renaissance man who
was a chemist, botanist, pharmacist, educator, and novelist who advanced
botanical chemistry and drug extraction. His personal library became the basis
of the Lloyd Library in Cincinnati which stands today as one of the great
repositories devoted to the history of botanicals. Lloyd helped them produce
safe medicines that were of pure quality.[20]

The Eclectics and their fellow travelers, known as the Reformed Practice,
continued to flourish until the early twentieth century. By 1900, there were
about 5,000 practicing Eclectic physicians representing about four percent of
all practicing physicians in the United States. The discovery of germ theory,
antiseptic surgery, general anesthesia, and other scientific advances in the end
overwhelmed the Eclectics and Reformed Practice. Their last medical school
established in Cincinnati in 1837 managed to survive Abraham Flexner's scrutiny in 1910, but closed its doors in 1929. Unlike traditional medical schools
of the nineteenth century, the Eclectic medical schools admitted women and
African-Americans as students as early as the 1840s.[21]

Homeopathy

While the Thomsonians and Eclectics had an impact on the development of
American medicine and pharmacy, homeopathy posed the most serious threat
to traditional medicine and had the widest following during the nineteenth
century. In part, it posed the greatest threat to traditional medicine because
it attracted a number of traditional physicians into its ranks. Homeopathy's
founder, Dr. Samuel Christian Friedrich Hahnemann (1755–1843), was born
in Meissen, Germany, studied at Leipzig and Vienna, and earned his medical degree from Erlangen. The young physician began practicing traditional
medicine, dutifully administering bleedings, blisterings, and purgings with
toxic chemicals to his patients and saw that these treatments were doing more
harm than good.[22]

He stopped practicing medicine and turned to translating medical and scientific texts in order to support his family. Hahnemann's epiphany came when he translated Dr. William Cullen's *Treatise on the Materia Medica* when the prominent Scottish physician argued that quinine works to treat malaria by strengthening the stomach. Hahnemann tested quinine's effects on himself, and soon learned that small doses of quinine induced fever in him. He then surmised that the quinine had provoked his immune system into healing itself, making him stronger and thus making him immune, or at least partially immune, to malaria. Hahnemann had discovered the prime principle of homeopathy, that the fever could cure fever or that like cures like (*similia similibus curentur*). Hahnemann had engaged in his first case of "drug-proving," a homeopathic method of testing drugs to see how the body reacts to them.[23] By 1796, Hahnemann had discovered the three founding principles of homeopathic treatment: (1) that like cures like, (2) diluted dosages must be used, and (3) that only one drug should be administered at a time. Hahnemann's disdain for poly-pharmaceuticals and his insistence on administering one drug at a time so that its effects could be monitored in the patient proved to be prescient. Hahnemann intuitively recognized, by taking detailed patient histories, that each patient was different and therefore required a personalized treatment plan. In this respect, Hahnemann and the homeopaths anticipated our own time's emphasis on personalized drug therapies and patient-centered care.[24]

In 1810, Hahnemann published the *Organon of the Medical Art* which formed the basis of homeopathic medicine, drawing it in sharp contrast with heroic medicine. The book became very popular, went through many editions, and remains in print today. In the book's 1833 preface Hahnemann observed,

> The adherents of the old school of medicine assail the body with large, often protracted and rapidly repeated doses of strong medicine, whose long-lasting, not infrequently terrible effects they do not know, and which they apparently make purposely unrecognizable through the commixture of more such unknown substances into one medicinal formula. The long-continued employment of such formulas inflicts new and, in part, ineradicable medicinal diseases upon the body.[25]

By contrast, Hahnemann reassured his patients that, "Homeopathy therefore *avoids anything that is even the slightest bit debilitating*. Homeopathy avoids as much as possible every arousal of pain because pain also robs the vitality."[26] The gentleness of homeopathic treatment compared to the harshness of heroic medical treatment proved to be homeopathy's primary attraction to patients. By the 1830s, when statistical and empirical medicine emerged, homeopathic practitioners were eager to use this new tool to provide evidence that patients who used homeopathic treatments fared much better during the recent cholera epidemics in the United States and had lower mortality rates than patients who underwent traditional treatment. As a result, in the United States life insurance companies granted discounts to policyholders who pledged to seek treatment only with homeopathic practitioners. In fact, exclusive homeopathic life insurance companies appeared in the 1840s.[27]

One of the most controversial aspects of homeopathy was its reliance on diluted doses of medicines, known as the law of infinitesimals. Hahnemann promoted this idea in his first volume of *Materia Medica Pura* which included 66 remedies. This would be followed by a further six volumes. According to Hahnemann, the more diluted a dose of medicine was the stronger it was. As homeopathic remedies came into increasing demand they were manufactured and sold in traditional pharmacies as well as in special homeopathic pharmacies. In 1876, *The Homeopathic Pharmacopoeia of the United States* was published and several editions appeared. By law it was recognized as official and had the same legal status as *The Pharmacopoeia of the United States* and *The National Formulary.*[29]

> The 1902 Sears, Roebuck and Company Catalogue offered a full line of homeopathic remedies including a 20-minute cure for the common cold![28]

Homeopathy was introduced in the United States by a Danish immigrant, Hans B. Gram, in New York in 1825. Gram was followed by the "father of American homeopathy," Constantine Hering, who opened the first homeopathic medical school in 1835 in Allentown, Pennsylvania, which was short-lived because all of its classes were conducted in German. Hering later founded another medical school in Philadelphia that later became Hahnemann University. In 1900, there were 22 homeopathic medical schools in the United States with over 15,000 practicing homeopaths, representing about one-sixth of all American physicians.[30] Still, homeopathy suffered a similar fate to the Thomsonians and the Eclectics, done in by germ theory, the Flexner Report, antiseptic surgery, general anesthesia, and the other scientific advances of the late nineteenth and early twentieth centuries. A few homeopathic practitioners, both physicians and pharmacists remain in practice today in the United States. By contrast, globally, homeopathy remains vital in Germany, the Netherlands, the United Kingdom, and has a large following in India, which has become a leader in homeopathic research and practice.

The Great American Medicine Show:
Patent Medicines and Nostrums

During the colonial era the British colonies in North America relied almost exclusively on the importation of herbs and medicines from Britain and continued to do so well into the American Revolution itself, in spite of the shortages that compelled them to improvise. In fact, when the British blockade nearly ended drug imports, many retailers simply refilled empty bottles with concoctions of their own and sold them under the old labels, with few consumers noticing any difference. With precious few actual apothecary shops in operation (see Chapter 9) most medicines were imported and

> The first patent bestowed upon a medicine by a British king was granted in 1698 to Epsom salts in England.[31] The first patent granted in British North America went to Thomas and Sybilla Masters of Philadelphia for refined corn known as Tuscarora Rice in 1715.[32]

sold as side-line merchandise by postmasters, goldsmiths, grocers, and tailors. Because patent medicines often consisted of large amounts of vegetable or fruit juice mixed with alcohol, opium, cocaine, or other narcotics, producers were not interested in patenting the shoddy ingredients of their medicines; they were keenly interested in gaining and maintaining the exclusive rights to their branding including labels, bottle shapes, bottle colors, and the slogans they used to advertise their brands. This proprietary era was first and foremost about branding and trademarking medicinal products.

Perhaps the first British patent drug advertised in an American newspaper was Daffy's Elixir Salutis, a laxative nostrum made from a tincture of senna, which appeared in the *Boston News-Letter*. The ad was placed by an apothecary named Nicholas Boone.[33] The elixir was named in honor of its creator, a British clergyman named Reverend Thomas Daffy, who introduced it in the mid-seventeenth century.[34] This newspaper was the first regularly published publication of its type in the British North American colonies. Originally it set out to earn its revenue by selling subscriptions, but its ownership soon learned that its livelihood depended on selling advertising. Coupled with the increasing popularity of nostrums newspaper advertising proved to be the perfect vehicle for promoting the use of these products. By 1800, there were 200 newspapers in the United States and by 1860 over 4,000.[35] With huge profits to be made, nostrum producers escalated the sensational claims they made for their products, which they did not have to prove. Claims to cure every condition from scrofula (tuberculosis of the lymph nodes in the neck) to cancer could be advertised with impunity. During the Civil War, the federal government placed a tax on patent medicines to finance the war effort. In 1859, nostrum sales in the United States topped $3.5 million and by 1904 nostrums earned $74.5 million.[36] With this amount of revenue, patent medicine producers formed a lobbying group called the Proprietary Association in 1881 which opposed attempts to regulate their nostrums. When the state legislature in North Dakota passed a medicine disclosure law the Proprietary Association had its members pull their advertising from all of the state's newspapers, inflicting a devastating loss of revenue for the newspapers

Lee's Bilious Pills was the first nostrum to be granted a patent in 1796
under the new U.S. Constitution.[37]

and causing a great furor. In
the next legislative session,
the legislature rescinded the
medicine disclosure law.[38]

Lydia Pinkham: The Queen of Patent Medicine

One of the most successful patent medicines of all time was born amid a family's dire financial circumstances. Lydia Estes Pinkham (1819–1883) was born
in Lynn, Massachusetts to a Quaker family that was involved in the aboli-
tionist, feminist, and temperance movements. Pinkham became intrigued by
alternative medical theories, especially those promoted by the Eclectics. She
learned to make home remedies, perhaps influenced by popular home medical reference books such as *Secret Nostrums and Systems* that contained recipes
for copying popular patent medicines. Similarly, the Philadelphia College
of Pharmacy published *Formulae for the Preparation of Eight Patent Medicines*
in 1824. She made several of these nostrums for her family and neighbors
to treat various health complaints. She married Isaac Pinkham and they had
three sons and a daughter. Isaac worked as a salesman but when the panic of
1873 hit, the family became destitute and when two men came to Lydia willing to pay cash for her nostrum, Lydia E. Pinkham's Vegetable Compound
was born. The family wondered if this product would sell in pharmacies. Her
son Dan distributed pamphlets around New York City, while Lydia prepared
her nostrum for sale. In one of advertising's greatest moments, her son Dan
had the idea of placing a picture of his mother's face on the label with her
signature. This revolutionized advertising because newspaper readers during
this era were not accustomed to seeing a woman's face in the newspaper.[39]
Lydia's reassuring matronly countenance gave the product a sense of legiti-
macy which, based on its ingredients, it did not deserve. The compound's
ingredients included black cohosh, unicorn root, life root, fenugreek seed,
pleurisy root, in a solution of 18 percent alcohol.[40]

The product sold moderately well until her son Will decided to have the
Boston Herald print the four-page pamphlet describing its benefits. Within five
years of placing this ad, the family sold $200,000 of Lydia's compound. They
bought a manufacturing plant where they gave tours emphasizing that theirs
was a family business. Lydia wrote a 62-page "Guide for Women" advis-
ing women about their health. Capitalizing on Victorian era modesty, Lydia
encouraged women to write to her for advice and until her death in 1883,
Lydia personally responded to her customers. One of her slogans reassured
women that men never see your letters. Lydia's compound remained a popular remedy until the family sold it to a pharmaceutical company that moved
the plant's operations to Puerto Rico. It can still be found today in specialty
pharmacies, more as a curiosity than cure.[41]

Patent medicines had great appeal to the public because they offered a
quick fix and made consumers feel as though they had some control over

their own treatment. Moreover, they contained alcohol and narcotics and provided a legitimate way for people, especially women, to consume them. Victorian era social norms, for all practical purposes, prohibited women from going to taverns lest they ruin their reputations. Ordering a patent medicine through the mail and consuming it at home provided a "respectable" way to drink or indulge in narcotics and, according to their wide appeal, many women did just that. The Sears, Roebuck, and Company mail order catalogue regularly featured a morphine-laced compound that housewives could slip into a wayward husband's coffee to keep him at home and away from the local tavern, which, based on brisk sales, undoubtedly must have worked.[42] From colonial times until the federal Pure Food and Drug Act of 1906, patent medicine producers could market their wares to the public without much government regulation, and did.

The Shakers: Bricks of Wholesome Herbs

Not all patent medicines were harmful or produced under shoddy conditions. Seeking religious freedom, the United Society of Believers in Christ's Second Appearance, better known as the Shakers, established their first settlement in 1774 in Niskeyuna and later at New Lebanon, New York. These Protestant sectarians who believed in male and female equality as well as celibacy were led by their founder, Mother Ann Lee. As sectarians they lived communally and were superb farmers, who by 1850 boasted an herb garden of 50 acres consisting of almost 200 native plants supplemented by over 30 more varieties imported from Europe. The Shakers proved to be an industrious and innovative people who parlayed their penchant for self-sufficiency into a thriving seed and herbal businesses that they started in 1794 and 1800, respectively. The Shakers began selling their "bricks" of carefully grown, harvested, and dried herbs to apothecaries in 1820. The first catalogues of their products appeared in 1830 and orders were shipped regularly overseas to England and France. The business thrived and according to George Bender, "by 1852, 42,000 pounds of roots, herbs, and barks were pressed, and 7,500 pounds of extracts were produced. By 1864, 16,450 pounds of extracts alone were produced."[43] By 1900, the Shakers' herb business had waned, but one product, Norwood's Tincture of Veratum Viride, remained popular for several decades. By 1947, the Shaker community in New Lebanon closed and by 1956 there were only 50 of them left living in three New England communities, victims of modernization.[44]

Patent Medicines That Became Today's Soft Drinks

Pharmacist John Pemberton and Coca-Cola

One of the developments that emerged from the patent medicine era was the invention of the soda fountain. The most famous soda, Coca-Cola, was invented in 1886 by a Confederate Civil War veteran and pharmacist named John Stith Pemberton (1831–1888) who worked at the Eagle Drug and Chemical Company in Columbus, Georgia. Pemberton had been wounded at the Battle of Columbus and searched for an elixir that could cure him of his morphine addiction. He experimented with coca wines such as the popular Vin Mariani and created Pemberton's French Wine Coca. When local authorities enacted temperance laws, Pemberton produced a non-alcoholic version of his French Wine Coca.[45] His bookkeeper, Frank Mason Robinson, designed the iconic name and logo for the new product that would become world renowned as Coca-Cola. The famous soda first sold for 5 cents per glass at Jacob's Pharmacy in Atlanta in May 1886, as a cure for headache and dyspepsia (upset stomach). In 1888, Pemberton sold the exclusive rights to his formula to an Atlanta pharmacist, Asa Griggs Candler, who began selling it in bottles in 1894. Candler's business acumen turned Coca-Cola into one of the best known products in the world through innovative advertising. When the United States entered World War II in 1941, the Coca-Cola Company went worldwide, opening bottling plants in every nation where American soldiers fought to provide American servicemen with a little piece of home. After the war, Coca-Cola kept these bottling plants open and, riding the wave of economic expansion, became one of the world's best known companies.[46]

Moxie

Another patent medicine that preceded Coca-Cola by a decade, began its career as Moxie Nerve Food, and today is sold mostly in New England as Moxie. Dr. Augustin Thompson of Union, Maine created the formula in 1876, selling it as a tonic to treat insomnia and nervousness. In 1884, Moxie was sold as a fountain drink as well as in bottles. Since 2005, Moxie has been designated as the official soft drink of the state of Maine and has remained a popular regional beverage in New England.[47]

In addition to aggressive newspaper advertising, in 1917 Thompson launched an ad campaign using traveling "horsemobiles," which were modified open touring cars that featured a driver who appeared to be riding a horse. His "horsemobiles" appeared at county fairs, parades, and community events across the nation. This campaign helped Moxie outsell Coca-Cola in 1920! Moxie's sales peaked in 1925 when, due to rising sugar prices, the company cut back on advertising, causing sales to plummet, rendering Moxie a regional favorite.[48]

Hires' Root Beer

At about the time Moxie was created, a Philadelphia pharmacist Charles Elmer Hires, developed his own formula for an American colonial folk drink, called root beer, made in part from sassafras. Hires claimed his drink would purify the blood and promote health. Hires sold his drink in packets of powder that cost 25 cents and could produce five gallons of root beer. In 1884, he produced syrup for soda fountains and when the company was incorporated in 1890, it sold its product in bottles. The Hires family sold the company in 1960 and after a number of corporate consolidations Hires' Root Beer gave way to Pepsi's Mug Root Beer in the 1990s.[49]

Dr. Pepper

Another patent medicine turned modern soft drink was Dr. Pepper, which made its debut in 1885. Charles Alderton, a Brooklyn born pharmacist working in Morrison's Old Corner Drug Store in Waco, Texas, was experimenting with various flavorings when he discovered the formula for a new drink which he offered to his boss, Wade Morrison, who agreed to sell it. The provenance of the name Dr. Pepper has stirred some debate. The most plausible explanation was that Wade Morrison named it in honor of a real life Dr. Pepper of Christiansburg, Virginia. Early in his career Morrison worked as a pharmacy clerk in that town and probably knew him. Despite the competition from the cola giants, Dr. Pepper has remained viable due to its clever advertising and unique flavor.[50]

Pepsi and 7 Up

Coca-Cola's main competitor, Pepsi, was introduced in 1898 in New Bern, North Carolina, by Caleb Bradham. First known as "Brad's Drink" it was renamed Pepsi Cola in 1903. It was first produced by Bradham in his home as a remedy for dyspepsia and as an energy drink. Pepsi's rivalry with Coca-Cola has become the stuff of American legend with the two cola giants controlling the world's soft drink market.[51]

In 1929, St. Louis businessman Charles Leiper Grigg introduced "Bib-Label Lithiated Lemon-lime Soda," which was later renamed 7 Up. The product contained lithium citrate which was based on a patent medicine used as a mood stabilizer. So, many of the patent medicines that had

Caleb Bradham invented Pepsi Cola in his North Carolina drugstore. By 1902, Bradham had enjoyed sufficient financial success from selling Pepsi that he closed his drugstore to work full time to develop the company in 1902. The company performed well until sugar prices fluctuated wildly after World War I, driving Bradham's company into bankruptcy in 1923. Bradham returned to working in a pharmacy and died in 1934.[52]

their roots as treatments for various ailments became the soft drinks we enjoy today and pharmacists played a key role in creating them.

Chapter Summary

American history runs in cycles of unbridled laissez-faire anything goes capitalism to calls for reform and government regulation to insure professional standards and public safety. In the history of American pharmacy, the era after the Revolution all the way up to the Pure Food and Drug Act of 1906, was one of these periods when anyone with an idea could develop, market, and distribute medicinal products to the public without any meaningful government regulation. To be sure, the publication of the first *Pharmacopoeia of the United States* in 1820, the establishment of the first American college of pharmacy in Philadelphia in 1821, and the Drug Importation Act of 1848 were attempts toward professionalization and regulation, but they proved to be no match for the unchecked growth and profitability of patent medicines. Without any safety standards, the public was indeed at risk and in many cases harmed, as evidenced by several prominent law suits. The implications of this spirit of laissez-faire proved to be both positive and negative. The positive feature was that new ideas and products were awaited by an eager market. The negative feature was the lack of safety or professional standards to protect the public from harmful products. The simple truth was that until the advent of germ theory in the 1870s, traditional medicine had little to offer patients. Medical and pharmacy education were uneven and haphazard, and in many cases the treatment was often worse than the disease, so it was not surprising that American patients opted for the convenience of self-medication until more effective forms of medical treatment were developed. History reminds us that humans will do anything to recover from illness and this aspect of human nature has not changed over the millennia. Despite the law books filled with government statutes regulating the practice of pharmacy, medicine, and drugs since the patent medicine era, a consumer can still find untested and unproven remedies in many pharmacies and department stores today.

Key Terms

William Buchan	Samuel Thomson	Albert Isaiah Coffin
John C. Gunn	*Lobelia inflate*	Wooster Beach
John Wesley	Alva Davis	The Eclectics

John Uri Lloyd	Lydia Pinkham	Charles Alderton
Samuel Hahnemann	the Shakers	Caleb Bradham
homeopathy	John Stith Pemberton	Charles Leiper Grigg
patent medicine	Augustin Thompson	

Chapter in Review

1 Describe the factors that contributed to the rise of alternative medicine in the United States.
2 Explain the theory of illness and treatments offered by the Thomsonians.
3 Explain the theory of illness and treatments offered by homeopathic medicine.
4 Account for the rise and popularity of patent medicines in the United States.
5 Describe the influence of religion on alternative medicine in colonial and nineteenth-century America.

Notes

1 Paul Starr, *The Social Transformation of American Medicine* (New York: Basic Books, 1982), 32.
2 Ibid., 33.
3 Ibid., 34.
4 Eunice Bonow Bardell, "Primitive Physick: John Wesley's Receipts." In *Pharmacy in History*, edited by Gregory J. Higby (Madison, WI: AIHP), Vol. 21 (1979), No. 3, 111–121.
5 John Uri Lloyd, ed., *The Life and Discoveries of Samuel Thomson* (Reprint, Cincinnati, OH: Bulletin of the Lloyd Library of Botany, Pharmacy, and Materia Medica, Bulletin No. 11, Reproduction Series No. 7, 1909), 69.
6 Ibid., 13.
7 Ibid., 18.
8 Ibid., 23.
9 John Uri Lloyd, ed., *The Life and Discoveries of Samuel Thomson*, 68.
10 Alex Berman and Michael A. Flannery, *America's Botanico-Medical Movements: Vox Populi* (New York: Pharmaceutical Products Press, 2001), 167–172.
11 David L. Cowen and William H. Helfand, *Pharmacy: An Illustrated History* (New York: Harry Abrams, 1990), 136.
12 James McWhorter, *Nature Cures: The History of Alternative Medicine in America* (New York: Oxford University Press, 2002), 26.
13 Gregory J. Higby, *In Service to American Pharmacy: The Professional Life of William Procter, Jr.* (Tuscaloosa: University of Alabama Press, 1992), 59–60.
14 John S. Haller, Jr., *The People's Doctors: Samuel Thomson and the Botanical Movement, 1790–1860* (Carbondale: University of Southern Illinois University Press, 2000), 29.
15 Metta Lou Henderson, *American Women Pharmacists* (New York: Pharmaceutical Products Press, 2002), 3.

16 David Armstrong and Elizabeth Metzger Armstrong, *The Great American Medicine Show* (New York: Prentice Hall, 1991), 26.

17 Ibid., 28.

18 John S. Haller, Jr., *The People's Doctors*, 239–240.

19 Ibid., 102–103.

20 David Armstrong and Elizabeth Metzger Armstrong, *The Great American Medicine Show*, 28–29.

21 Ibid., 29.

22 John S. Haller, Jr., *The History of American Homeopathy* (New York: Pharmaceutical Products Press, 2005), 9–10.

23 Ibid., 11. Barbara Griggs, *Green Pharmacy* (Rochester, VT: Healing Arts Press, 1997), 171–172.

24 Wenda Brewster O'Reilly, ed., *The Organon of the Medical Art by Dr. Samuel Hahnemann* (Redmond, Washington: Birdcage Books, 1996), xvi–xvii.

25 Ibid., 2.

26 Ibid., 4.

27 John S. Haller, Jr., *The History of American Homeopathy*, 114–115.

28 Sears, Roebuck, and Company, *The Sears, Roebuck Catalogue 1902 Edition* (Reprint: New York Crown Publishers, 1969), 451.

29 David L. Cowen and William H. Helfand, *Pharmacy*, 135.

30 David Armstrong and Elizabeth Metzger Armstrong, *The Great American Medicine Show*, 35.

31 David L. Cowen and William H. Helfand, *Pharmacy*, 167.

32 Metta Lou Henderson, *American Women Pharmacists*, 2.

33 George Griffenhagen and James Harvey Young, "Old English Patent Medicines in America." In *Contributions from the Museum of History and Technology* (Washington, DC: Smithsonian Institution, 1959; Reprint, Madison, WI: AIHP), Vol. 34 (1992), No. 4, 205–206.

34 A.C. Wootton, *Chronicles of Pharmacy, Vol. 2* (Boston, MA: Milford House, 1972), 172.

35 James Harvey Young, *The Toadstool Millionaires* (Princeton, NJ: Princeton University Press, 1961), 39.

36 Ibid., 110.

37 James Harvey Young, *American Self-Dosage Medicines: An Historical Perspective* (Lawrence, KS: Coronado Press, 1974), 2–3.

38 James Harvey Young, *The Toadstool Millionaires* (Princeton, NJ: Princeton University Press, 1961), 211.

39 Paul Starr, *The Social Transformation of American Medicine*, 164–165. William H. Helfand, *"Let Us Sing of Lydia Pinkham" and Other Proprietary Medicines* (Madison, WI: AIHP, 1994), 21–24.

40 Vince Staten, *Do Pharmacists Sell Farms?* (New York: Simon & Schuster, 1998), 174.

41 David Armstrong and Elizabeth Metzger Armstrong, *The Great American Medicine Show*, 165.

42 Sears, Roebuck, and Company, *The Sears, Roebuck Catalogue 1902 Edition*, 450.

43 George Bender, *Great Moments in Pharmacy* (Detroit, MI: Northwood Institute Press, 1966), 114.

44 Ibid., 117. Thomas D. Clark and F. Gerald Ham, *Pleasant Hill and Its Shakers* (Pleasant Hill, KY: Sherktown Press, 1968), 1–87.

45 Monroe Martin King, "Dr. John S. Pemberton: Originator of Coca-Cola," *Pharmacy in History*, Vol. 29 (1987), No. 2, 85–87.

46 Anne Cooper Funderburg, *Sundae Best: A History of Soda Fountains* (Bowling Green, OH: Bowling Green University Popular Press, 2002), 73.

47 Ibid., 67–69.

48 Ibid., 69.

49 Ibid., 92–93.

50 Ibid., 72–74.

51 Ibid., 78–81.

52 Ibid., 79.

11 American Pharmacy Organizes and Promotes a New Vision for the Practice of Pharmacy

Why did pharmacy organizations emerge and how did they shape the course of American pharmacy?

> Pharmacy in the United States was organized from the top down, not from the bottom up as one might expect.[1]
>
> (George Griffenhagen)

As the proliferation of sectarian medical systems emerged and the sales of patent medicines flourished during the nineteenth century in the United States, there were groups of concerned physicians, pharmacists, and others who dedicated their efforts to promoting drug safety and professional standards. Building on the colonial and American Revolution experience, professionalization standards were set by local governments and Congress, local groups formed to discuss the state of medical and pharmacy practice. These discussions led to a series of landmark events in the history of American pharmacy including the creation of the first *Pharmacopoeia of the United States* in 1820, the founding of the Philadelphia College of Apothecaries in 1821, the founding of the American Medical Association in 1847, the passage of the Drug Importation Act of 1848, the founding of the American Pharmaceutical Association in 1852, and ultimately the passage of the Pure Food and Drug Act in 1906. These events also happened during a time of dramatic changes that brought medicine and pharmacy into the modern era.

The Rise of National Pharmacopoeias and the *United States Pharmacopoeia*

The American and French Revolutions of the eighteenth century gave birth to the modern nation-state and an even more powerful force known as nationalism. The rise of newspapers, railroads, steamboats, and other innovations provided the impetus for local and regional interests to look toward the nation-state as the legitimizing force in all matters. For physicians and pharmacists, this meant organizing on a national level to achieve the degree of professional recognition they sought. Since the Middle Ages, apothecaries had been working from local and regional formularies and dispensaries that were quite practical, but lacked standardization. The first call for a national pharmacopoeia in the United States came from Dr. John Morgan, who proposed a pharmacopoeia for the Commonwealth of Pennsylvania in 1787.[2] Soon state medical societies from New York, Delaware, Connecticut, South Carolina, and Massachusetts saw the value of a national pharmacopoeia. The Massachusetts Medical Society publicly voiced its concern over the lack of drug safety to the commonwealth's legislature in 1786. They drafted a pharmacopoeia in 1807 for comment and the next year, a 272-page pharmacopoeia listing 536 drugs, was adopted by practitioners in the commonwealth.[3] Several state medical societies recognized the value of a national pharmacopoeia and, led by a call from Dr. Lyman Spalding (1775–1821) of New York, a convention was held in the Senate chamber on January 1, 1820, in Washington, DC. Delegates from the Northern, Middle, Southern, and Western states attended and they were expected to bring their own pharmacopoeias to the meeting. The Massachusetts, New York, and Pennsylvania delegates had the most influence with nearly 90 percent of the *Massachusetts Pharmacopoeia* being incorporated into the national pharmacopoeia. The national pharmacopoeia was published in Boston on December 15, 1820.[4] The book listed drugs and their preparation that appeared in both Latin and English. The authors of the book argued that the use of Latin would render it "more intelligible to foreigners, and more useful in those districts of the United States where the French and German languages continue to be spoken."[5] The authors also had the foresight to suggest that a revision of the *United States Pharmacopoeia* (*USP*) be undertaken once every ten years. While the first national pharmacopoeia was produced almost entirely by physicians, by 1877 pharmacists would be sufficiently organized as a profession to oversee its subsequent revisions, largely due to the leadership of Drs. Edward Squibb (1819–1900) and Charles Rice (1841–1901). By 1882, pharmacists became the guiding force of subsequent editions of the *USP*.[7]

In a strange twist of fate, shortly after the December 15, 1820 publication of the first *Pharmacopoeia of the United States of America*, Lyman Spalding was involved in a tragic freak accident. Spalding had been walking along Pearl Street in New York, when he was struck on the head by a box of rubbish that fell on him from a second story window. The blow was cushioned somewhat by his hat and wig, but sadly left him incapacitated. He left New York to live in his home town of Portsmouth, New Hampshire, where he died on October 21, 1821, but not before he became known as the driving force in founding the *Pharmacopoeia of the United States of America*.[6]

Having taken charge of the *USP*, pharmacists under the auspices of the American Pharmaceutical Association, which was founded in 1852, published a complementary book called the *National Formulary of Unofficial Preparations* in July 1888. The *National Formulary* was based on the work of a New York pharmacist named Samuel Bendiner who inspired the creation of the *New York and Brooklyn Formulary*. Charles Rice, a hospital pharmacist from New York, saw the benefit of Bendiner's work as a way to encourage physicians to prescribe drugs from this formulary rather than prescribe more risky proprietary patent medicines.[8] The *National Formulary* contained drugs that physicians often prescribed, but were not listed in the *USP*. It also contained new drugs that often were incorporated into subsequent revisions of the *USP*. The *National Formulary* had proven its value and it became official with the passage of the Pure Food and Drug Act of June 30, 1906. Since 1975, both have been published in a single volume with the pharmacopoeia containing drug substances and dosages; whereas the formulary contained drug ingredients.[9]

The Origins of Pharmacy Education in the United States

With the rapid advances in chemistry and the growing complexity of preparing drugs the need for specially trained pharmacists became apparent. In Britain, France, and the German states, pharmacy education became increasingly institutionalized combining didactic education in the natural sciences and some law with varying lengths of apprenticeship, followed by a licensure examination. Pharmacy education in Europe was often tied to medical education with practical courses in chemistry and later laboratory work. In the European system, physicians dominated pharmacy education. By contrast, the education scene for pharmacists in the United States was haphazard at best. Even in the larger cities such as Philadelphia, self-proclaimed charlatans often competed with highly skilled and conscientious apothecaries for business. To address this unfortunate situation, Dr. John Redman Coxe, the dean of the medical faculty at the University of Pennsylvania, proposed establishing a Master of Pharmacy degree that would be open to students as well as some "deserving" apothecaries operating in Philadelphia. The "deserving" apothecaries would be selected by the medical faculty at the university; others who were deemed "neglectful or indifferent" would not receive recognition. As soon as this proposal was published in February 1821, it created great controversy among the city's apothecaries. Their professional pride and individual dignity were clearly at stake, although at least 16 of the city's apothecaries signed on to Coxe's proposal.[10]

The city's apothecaries met at Carpenter's Hall on February 23, 1821, to discuss the proposal and to develop their collective response. Some supported Dr. Coxe's plan, but most favored rejecting the plan and plotting an independent course that would be shaped by and for apothecaries.

At a second meeting held in March 1821, the committee proposed a College (association) of Apothecaries to monitor the quality of drugs being sold and to establish a school of pharmacy. They offered a constitution for the association that called for the founding of a 16-member board of trustees overseeing quarterly meetings. Membership dues were $5 per year and membership was open to practicing apothecaries and those who would earn diplomas from the College. The constitution charged the board with founding a school of pharmacy with a library and hiring lecturers to teach *materia medica* (the natural history of drugs), pharmacy, and chemistry. The plan also called for a committee of pharmacists to be appointed whose charge would be to inspect all drugs in the market to protect the public from harmful products. This proposal was major step toward professional autonomy, since physicians traditionally were expected to perform pharmacy inspections. A committee of equity would be established to settle any disputes among the membership, usually involving business practices.[11]

The first meeting of the Philadelphia College of Apothecaries (its name changed to the Philadelphia College of Pharmacy in 1822 when it received its charter) occurred on March 27, 1821. Charles Marshall, the son of the legendary pharmacist Christopher Marshall, became the College's first president and served until he retired in 1824. He was succeeded by William Lehman who, in turn, was succeeded in 1829 by Daniel B. Smith (1792–1883) who served as president for the next 25 years. The first lectures were held on November 9, 1821, in a rented hall, by Dr. Samuel Jackson who delivered courses on *materia medica* and pharmacy, and cost students $12. Jackson taught for six more years and was succeeded by Dr. Benjamin Ellis. Dr. Gerard Troost began his chemistry course the next evening, which cost students $10. In order to graduate students were required to take two courses from each of the subjects offered. Troost taught for only one year and was succeeded by Dr. George Wood who, in 1831, switched to teaching *materia medica* and pharmacy. Dr. Franklin Bache became the new chair of chemistry and these two men pioneered a program that set the standard for pharmacy education in the United States, yielding outstanding graduates including William Procter, Jr., Edward Squibb, John Maisch, Joseph Remington, Linwood Tice, Charles Lawall, and others.[12] Another extension of the College's influence was the founding of the *Journal of the Philadelphia College of Pharmacy* in 1825, the first pharmacy journal in the English language. In 1835, it was renamed the *American Journal of Pharmacy*, the oldest journal exclusively dedicated to pharmacy issues. William Procter, Jr. became the journal's editor in 1850 and for the next two decades helped shape the course of pharmacy in all of its aspects in its struggle to become recognized as a profession with a business side as opposed to a business with a professional side.[13]

Although the Philadelphia College of Pharmacy established a new paradigm for pharmacy education in America, most practitioners still believed that pharmacy was best learned through apprenticeships and the early proprietary

Table 11.1 Early Pharmacy Schools Founded by Pharmaceutical Associations

1821	Philadelphia College of Apothecaries
1822	Philadelphia College of Pharmacy
1823	Massachusetts College of Pharmacy
1829	College of Pharmacy of the City of New York
1840	Maryland College of Pharmacy
1850	Cincinnati College of Pharmacy
1859	Chicago College of Pharmacy
1864	St. Louis College of Pharmacy (precursors in 1854 and 1857)[14]

schools of pharmacy struggled. Typically, in its first 20 years of operation the Philadelphia College of Pharmacy graduated an average of 21 students per year. Philadelphia was followed by the founding of the Massachusetts College of Pharmacy in Boston in 1823. Massachusetts offered occasional evening lectures and discussion sessions until 1867, when it offered a regular sustained course of evening instruction. While Philadelphia had consistent enrollment, it only began to offer laboratory instruction in 1870. By 1865, there were six schools offering degrees in pharmacy and the Tulane Medical School which offered a course in pharmacy.[15] Still by 1900, there were more than 60 schools offering pharmacy programs.

The Founding of the American Pharmaceutical Association

With a laissez-faire market in place, one of the major challenges for both physicians and pharmacists was drug safety. Since colonial times, the American colonies proved to be a lucrative dumping ground for weak, expired, or adulterated drugs from Europe. This state of affairs continued after the Revolution well into the nineteenth century. The situation became acute in New York during the 1840s when an apothecary, Ewen McIntyre, discovered that a shipment from Britain labeled as calcium carbonate was actually calcium sulfate. McIntyre's boss referred the matter to the New York College of Pharmacy for investigation. Examiners discovered that other products had also been adulterated or mislabeled, posing a significant danger to the public. A letter of protest was sent to the British manufacturer who replied that this was a matter of what the American market would bear.[16] Rumors of adulterated drugs causing harm to American soldiers fighting in the Mexican–American War also played a role in prompting a hesitant Congress to take action.[17]

As a result of these cases, pharmacists and physicians nationwide appealed to the Congress for legislation, resulting in the Drug Importation Act of 1848. This was an initial trial for the newly formed American Medical Association (AMA). The AMA had formed in Philadelphia on May 7, 1847, in part, according to its charter, as an "Influence for greater scientific accuracy and for more dependable therapeutic agents."[18] The law specified that

Figure 11.1 William Procter, Jr. (1817–1874) perhaps more than any other figure became known as the "the father of American Pharmacy" for his role in founding and nurturing the APhA and for serving for 22 years as the editor of the *American Journal of Pharmacy*.

Source: Courtesy of the National Library of Medicine.

special examiners be hired to inspect all medicinal products for their fitness using the *USP* and other dispensatories as standards. Examiners were hired at the six busiest ports in the United States: New York, Boston, Philadelphia, Baltimore, Charleston, and New Orleans. Unfortunately, shoddy imported products still appeared, which raised calls for better training for the inspectors. Another factor was that many of these port inspectors were political appointees in a system where political loyalty trumped professional competence. Unscrupulous drug exporters simply evaded the law by reshipping drugs rejected by one port to another port where they often were approved for sale.[19]

To address this problem, the New York College of Pharmacy invited delegates from the other colleges to meet in New York to develop a training program for these inspectors. One delegate from the Philadelphia College of Pharmacy, named William Procter, Jr. (1817–1874), saw the opportunity to parlay this meeting into a call to form a national association for pharmacists. Procter was a visionary who had admired the Pharmaceutical Society of Britain that had been founded in 1841, and wanted a similar organization to advance pharmacy as a profession in the United States.[20] The delegates met in New York in October 1851 and agreed on a set of training standards for the drug inspectors and agreed to meet in Philadelphia in 1852 to discuss the formation of what would become the American Pharmaceutical Association (APhA).[21]

About 20 delegates from various states met at the Main Hall of the Philadelphia College of Pharmacy from October 6–9, 1852 to form an umbrella organization that would advance the cause of its membership and improve communication among the nation's pharmacists. The delegates discussed many important issues and charted the course of this new organization. Nine important objectives were presented to the delegates, which had been developed in 1851 by a committee chaired by William Procter, Jr.:

1   Establish a national organization with a constitution and code of ethics.
2   Promote pharmacy schools and support education for its members.
3   Increase the selection standards for apprentices and improve their training.
4   Investigate patent medicines and cases of quackery.
5   Strengthen federal and state laws concerning the inspection of imported drugs.
6   Adopt a national pharmacopoeia as a guide in preparing medicines.
7   Limit the indiscriminate sale of poisons.
8   Promote the separation of pharmacy from medicine: prevent physicians from operating pharmacies and prevent pharmacists from practicing medicine.
9   Promote the presentation of original research on pharmacy and science.

The delegates supported the objectives, but debated whether membership should be open to every individual pharmacist or be representational. They wisely opted for individual membership. With that debate resolved the delegates voted and adopted a constitution for the new organization and also supported a Code of Ethics for "all pharmaceutists and druggists." The Code of Ethics was already familiar to the delegates from the Philadelphia College of Pharmacy, since it was the Code that they had lived by as faculty members of the College since 1848. Still, for the larger body of delegates the Code of Ethics became a source of controversy early in the APhA's history as it called upon its members to shun the sale of patent medicines that were of dubious quality. Edward Parrish (1822–1872), a friend and colleague of William Procter at the Philadelphia College of Pharmacy, worried that this clause would exclude many potential members who relied upon the sale of these shoddy remedies in order to stay in business in an increasingly competitive market. In deference to building the membership of the fledgling organization, the APhA dropped the statement on ethics in 1857. Nonetheless, over the next several decades, debates about reviving it remained and, due in large part to the efforts of Charles Lawall, in 1922 the APhA adopted a new Code of Ethics.[22]

Daniel B. Smith, the long-time President of the Philadelphia College of Pharmacy, was selected as the APhA's first president. The ubiquitous William Procter, Jr. became the organization's Corresponding Secretary serving in that capacity for the next two decades, becoming the conscience and guiding force of not only the APhA, but for American pharmacy in general, earning him the title of the "Father of American Pharmacy." In addition to his key role in the APhA, Procter served as the long-standing editor of the *American Journal of Pharmacy* which had a profound influence in shaping the course of pharmacy in the United States. A remarkable figure in the history of American pharmacy, Procter's career, which began in the 1840s and ended with his death in 1874, made him the critical link between the disorganization of pharmacy during the Early Republic to the post-Civil War era, which ushered in the establishment of organizations that promoted the professionalization of pharmacists. Procter's organizational vision, which was shared by the founders of the APhA, was to create a national umbrella organization that would foster the growth of other national, state, and local pharmacy organizations that would serve the needs of various constituencies within pharmacy.[23]

Pharmacy Education

As the APhA's membership grew in the decades following its founding it soon became apparent that its broad-based approach to promoting the professional and scientific aspects of pharmacy had limits. Even in the latter half of the nineteenth century, American pharmacists were a diverse lot involved in manufacturing, wholesaling, retail dispensing, education, and hospital pharmacy. Squabbles among these interest groups threatened to overwhelm the APhA. Moreover, there were other divisions that emerged. Some pharmacists owned their own shops and were known as "druggists" while others who were employees of these wholesale and retail drugstores were known as "drug clerks." By the 1870s, disputes between capital and labor emerged. This division of labor and degree of specialization begged for the establishment of specialized organizations that could devote themselves directly to the interests important to their members. The APhA played a key role in fostering these organizations by creating subsections or interest groups to address the varied interests of its members. Amid this atmosphere several key specialized pharmacy organizations were founded. The first of these organizations was the Conference of Teaching Colleges of Pharmacy represented by delegates from the "teaching colleges" of Chicago, Maryland, Massachusetts, New York, and Philadelphia and from the New Jersey Pharmaceutical Association in 1870 under the auspices of the annual APhA meeting. The Conference began by excluding the University of Michigan's delegation because it awarded pharmacy degrees based exclusively on didactic work with no requirement for apprenticeship, which the teaching schools viewed as unacceptable. The Conference labored on for 13 years without accomplishing much, largely because it had no real authority other than to recommend policy to the member colleges.[24]

The successor to the ill-fated Conference of Teaching Colleges of Pharmacy was the Conference of Pharmaceutical Faculties, founded in 1900. Signaling a major shift in the direction of pharmacy education, Albert B. Prescott (1832–1905), the founding dean of the University of Michigan's Pharmacy School, was elected as the president of the first conference.[25] This organization represented 21 schools of pharmacy in the United States in 1900. In 1925, the Conference was renamed the American Association of Colleges of Pharmacy (AACP) and directed much of its effort toward establishing national standards in pharmacy education. Prior to an official accreditation system, AACP was able to gain compliance with the standards it set by denying admission to any school that did not agree to comply with its standards. In 1937, AACP established the *American Journal of Pharmaceutical Education*, one of the most influential and respected journals of its kind in pharmacy.[26] Today, AACP has over 130 member schools of pharmacy and remains at the forefront of innovation in pharmacy education.

The National Association of Boards of Pharmacy

Another key organization born out of the APhA Section on Education and Legislation was the Association of Boards of Pharmacy and Secretaries of State Pharmaceutical Associations in the early 1890s. This ill-fated organization lasted until 1904, when it split into two organizations: the National Association of Boards of Pharmacy (NABP) established in 1904 and the Conference of Pharmaceutical Secretaries in 1927. The latter organization, renamed in 1949, became the National Council of State Pharmaceutical Association Executives.[27] The NABP, founded by 17 state boards, worked with state boards of pharmacy to promote minimum standards of competence for pharmacy licensure, although the states retained their own legislative autonomy. Prior to licensure in any state, aspiring pharmacy students must take the NABPLEX examination and achieve a satisfactory score in order to earn their license to practice.[28] The same is true on the Multistate Pharmacy Jurisprudence Examination. Moreover, the NABP has worked to make the reciprocal licensing of pharmacists who wish to practice in another state a more streamlined process.[29] The NABP was also instrumental in creating an official accreditation body responsible for maintaining national standards in pharmacy education, known as the American Council on Pharmaceutical Education (ACPE) which was founded in 1932 and was later renamed the Accreditation Council on Pharmaceutical Education. It is now known as the Accreditation Council for Pharmacy Education and uses the same acronym.

Commercial Pharmacy Organizations

In order to address the commercial aspects of pharmacists who owned their own businesses, the National Retail Druggists Association was founded in 1883, but appeared doomed from the start because it tried to cater to both wholesalers and retailers and dissolved in 1887. Patent medicines and the price cutting of drugs caused the APhA to establish a Section on Commercial Interests. Disagreements abounded with each annual meeting until 1898, when led by the efforts of the Chicago Retail Druggists Association and the leadership of Joseph Price Remington (1847–1918), the National Association of Retail Druggists (NARD) was founded. NARD was renamed the National Community Pharmacists Association (NCPA) in 1996. A number of joint meetings were held by the APhA and NARD over the decades, but cooperation has remained elusive.[30]

Manufacturers' Organizations

Especially after the Civil War the manufacturing of drugs replaced the need for pharmacists to compound their own prescriptions and ushered in a new era of pharmacy practice.[31] The American Association of Pharmaceutical Chemists was founded in 1908 and later merged with the National Association of Manufacturers of Medicinal Products, founded in 1912, to form the Pharmaceutical Manufacturers Association in 1958. This organization later became the Pharmaceutical Research and Manufacturers of America.[32] When injectable drugs came onto the market the Parenteral Drug Association was founded in 1946 to deal with the specialized requirements involved in producing these drugs.

Wholesalers' Organizations

The first organization to represent the interests of wholesale pharmacists was the Western Wholesale Druggists Association, founded in 1876. In 1900, this organization became the National Wholesale Druggists Association (NWDA). In order to compete with price cutting and underselling many retail pharmacy owners banded together to form "buying clubs" that allowed them to take advantage of volume discounts from manufacturers and wholesale distributors. The first organization of this type was formed in 1905 as the American Druggists Syndicate, and this was followed by the founding of the Associated Drug Companies of America in 1906. In 1915, the Federal Wholesale Druggists' Association (FWDA) replaced these two organizations. Eventually the FWDA merged with the NWDA and in 2001 NWDA became the Healthcare Distribution Management Association.[33]

Chain Pharmacies

By the 1890s, a new phenomenon began to take root in commercial pharmacy: the chain drug store, a place that offered medicine with the convenience of a general store. As a function of efficiency, volume selling, and customer convenience the phenomenon of the chain drugstore emerged in the early twentieth century and by the 1920s had become a mainstay in retail pharmacy. While only 6 percent of all pharmacies in the United States were chain drugstores during the 1920s, they had unique interests apart from what NARD and other pharmacy organizations could offer them. The National Association of Chain Drug Stores (NACDS) was established in 1933 by Wallace J. Smith of Read Drug and Chemical Company.[34]

Hospital Pharmacists

While hospitals date back to the Pennsylvania Hospital in 1751, the modern American hospital was born in the 1920s as innovative institutions where patients expected to get well. In 1927, Harvey A.K. Whitney (1894–1957) pioneered the first hospital pharmacy internship program at the University of Michigan Hospital that set the precedent for today's residency programs.[35] Similarly, a year later at the University of Iowa Hospital, staff pharmacists went on clinical rounds with physicians as members of a health care team. Nonetheless, starting out as a subsection of the APhA's Section on Practical Pharmacy and Dispensing in 1936, hospital pharmacists remained a minority in the profession whose interests were overshadowed by those of community pharmacists. In 1942, the American Society of Hospital Pharmacists (ASHP) was founded with H.A.K. Whitney, Sr. serving as its first president. ASHP founded the *American Journal of Hospital Pharmacy* in 1943 and the *American Hospital Formulary Service* in 1959 with Donald E. Francke (1910–1978) serving as the editor of both publications. ASHP later changed its name in 1994 to the American Society of Health-System Pharmacists.

Clinical Pharmacists

With the rise of the modern American hospital during the 1920s, advocates such as John C. Krantz, Jr. called for clinical services to become a part of pharmacists' care of patients.[36] After World War II, especially with the introduction of antibiotics and sulfa drugs, there were calls for clinical services in both hospital and community pharmacy settings. Nonetheless, clinical pharmacy came into its own in the late 1960s with ASHP and APhA introducing the idea of "patient-oriented care." The Ninth Floor Pharmaceutical Service Project conducted at the University of California, San Francisco in 1966 physically placed pharmacists next to the general surgery nursing station on the 9th floor of Moffitt Hospital to provide direct clinical services to patients as an integral part of the health care team.[37] Born out of the APhA's Clinical Practice Section, which was established in 1975, the American College of Clinical Pharmacy (ACCP) was founded in Kansas City in 1979 as a professional and scientific society to enhance the practice of clinical pharmacy.[38] ACCP worked to get pharmacotherapy recognized by the Board of Pharmaceutical Specialties (now known as the Board of Pharmacy Specialties) as a specialty within pharmacy and succeeded in 1988.[39]

Chapter Summary

While some progress had been made to organize pharmacy formally during the American Revolution, largely to support the war effort, much of this progress lapsed after the war ended and Adam Smith's "invisible hand" of free market economics dominated the American scene, especially pharmacy. Unlike the Europeans, American apothecaries had little experience with guilds other than adopting the apprenticeship system for training new apothecaries. This, coupled with the geographic vastness of this new land, complicated the task of organizing American pharmacy formally. Nonetheless, in the early nineteenth century, progress in organizing pharmacy began with the creation of the first *United States Pharmacopoeia* in 1820, followed by the establishment of the Philadelphia College of Apothecaries in 1821. As the young nation grew, safety concerns about the quality of imported drugs led to the Drug Importation Act of 1848, which in turn led to the founding of the American Pharmaceutical Association (APhA) in 1852. Under the auspices of the APhA a host of other professional, scientific, regulatory, educational, and business organizations emerged to serve the increasingly divergent interests of pharmacy. With the founding of each new pharmacy organization, the vision of the APhA's founders such as William Procter, Jr. have come to fruition, advancing the pharmacy profession in every conceivable field of endeavor.

Key Terms

Pharmacopoeia of the
United States of America

Philadelphia College of
Pharmacy

American Medical
Association (AMA)

American Pharmacists
Association (APhA)

Drug Importation Act
of 1848

Lyman Spalding

National Formulary

John Redman Coxe

Charles Marshall

William Procter, Jr.

Charles Lawall

Alfred B. Prescott

National Association
of Boards of Pharmacy
(NABP)

National Community
Pharmacists
Association
(NCPA)

Pharmaceutical
Research and
Manufacturers of
America (PhRMA)

Health Care
Distribution
Management
Association

National Association
of Chain Drug Stores
(NACDS)

American Society
of Health-System
Pharmacists (ASHP)

American College of
Clinical Pharmacists
(ACCP)

Chapter in Review

1 Identify the obstacles that hindered the organization of American pharmacy.
2 Trace the origin of the *Pharmacopoeia of the United States of America*.
3 Trace the origins of the Philadelphia College of Pharmacy and the subsequent role it played in shaping the course of American pharmacy.
4 Explain the significance of the Drug Importation Act of 1848.
5 Trace the origins of the APhA and its goals for the pharmacy profession.
6 Explain why there are so many pharmacy organizations.

Notes

1 George Griffenhagen, "Pharmacy Organizations, 1852–1902." In *American Pharmacy: A Collection of Historical Essays*, edited by Gregory Higby and Elaine Stroud (Madison, WI: AIHP, 2005), 87.

2 Glenn Sonnedecker, comp., *Kremers and Urdang's History of Pharmacy* (4th ed., Madison, WI: AIHP, 1976), 255.

3 Glenn Sonnedecker, "The Founding of the U.S. Pharmacopoeia." In *Pharmacy in History*, edited by Gregory Higby, Vol. 36 (1994), No. 1, 9.

4 Lee Anderson and Gregory Higby, *The Spirit of Voluntarism: A Legacy of Commitment and Contribution* (Rockville, MD: The United States Pharmacopoeial Convention (USPC), 1995), 25.

5 Samuel L. Mitchill, ed., *Pharmacopoeia of the United States* (Boston, MA: Wells and Lilly, 1820). Reprint (Madison, WI: AIHP, 2005), 26.

6 Lee Anderson and Gregory Higby, *The Spirit of Voluntarism*, 25.

7 David L. Cowen and William H. Helfand, *Pharmacy: An Illustrated History* (New York: Harry Abrams, 1990), 138.

8 George Griffenhagen ed., *150 Years of Caring* (Washington, DC: American Pharmaceutical Assocation, 2002), 111.

9 David L. Cowen and William H. Helfand, *Pharmacy*, 142.

10 George Bender, *Great Moments in Pharmacy* (Detroit, MI: Northwood Institute Press, 1966), 105–106.

11 Ibid., 108.

12 Ibid., 109.

13 Gregory Higby, *In Service to American Pharmacy: The Professional Life of William Procter, Jr.* (Tuscaloosa: University of Alabama Press, 1992), 105.

14 Glenn Sonnedecker, *Kremers and Urdang's History of Pharmacy*, 193.

15 Ibid., 231.

16 Ibid., 198.

17 George Griffenhagen, *150 Years of Caring*, 5.

18 George Bender, *Great Moments in Medicine* (Detroit, MI: Northwood Institute Press, 1965), 218.

19 George Griffenhagen, *150 Years of Caring*, 130.

20 Gregory Higby, *In Service to American Pharmacy*, 155.

21 Gregory Higby ed., *American Pharmacy: A Collection of Historical Essays* (Madison, WI: AIHP, 2005), xi.

22 George Bender, *Great Moments in Pharmacy*, 122.

23 Gregory Higby, *In Service to American Pharmacy*, passim.

24 George Griffenhagen, *150 Years of Caring*, 121–122.

25 Glenn Sonnedecker, *Kremers and Urdang's History of Pharmacy*, 249.

26 Ibid., 250.

27 Ibid., 205.

28 Ibid., 218.

29 George Griffenhagen, "Pharmacy Organizations, 1902–1952," 94.

30 C. Fred Williams, *A Century of Service and Beyond* (Alexandria, VA: National Community Pharmacists Association, 1998), passim. Gregory Higby, "From Compounding to Caring: An Abridged History of American Pharmacy." In *Pharmaceutical Care*, edited by Calvin Knowlton and Richard Penna (2nd ed., Bethesda, MD: ASHP, 2003), 31–32.

31 Ibid. (Higby), 30.

32 George Griffenhagen, "Pharmacy Organizations, 1902–1952," 97.

33 Ibid., 97–98. Jane Mobley, *Prescription for Success: The Chain Drug Story* (Kansas City, MO: The Lowell Press, 1990), 18.

34 Ibid., 96–97.

35 Robert Elenbaas and Dennis Worthen, *Clinical Pharmacy in the United States: Transformation of a Profession* (Lenexa, KS: American College of Clinical Pharmacy (ACCP), 2009), 15.

36 Ibid., 14.

37 Ibid., 35.

38 Ibid., 55.

39 Ibid., 76.

12 Learning How to Learn—From Apprenticeship to the Doctor of Pharmacy (Pharm.D.)

The Struggle for Professional Recognition

How did pharmacy education change over the centuries? How did pharmacy education shape the pharmacy profession?

Since medieval times, apothecaries learned their art at the side of a preceptor while serving various terms of apprenticeship in hopes of becoming a master in a European guild and opening their own shops. With the rapid scientific advancements of the late eighteenth and early nineteenth centuries, especially in alkaloid chemistry, several European nations instituted formal programs of instruction for apothecaries usually in conjunction with medical educators. The German states influenced by the traditions of Frederick II's *Constitutiones* (*c.*1240) developed private pharmaceutical teaching institutes in the late eighteenth century that taught subjects in the basic sciences to supplement the practical experience gained in apprenticeships.[1]

The Roots of Modern Pharmacy Education in Prussia

By the early nineteenth century, several German universities and polytechnics offered courses to aspiring pharmacy students in order to help them prepare for state examinations, which were the most rigorous in Europe at this time. For example, Prussian law required that candidates for examinations serve a five-year regular apprenticeship or serve for three years as an assistant in a licensed shop. In either case, candidates were also required to complete two full courses of botany, chemistry, natural history (*materia medica*), pharmacy, and medical law. The board of examiners consisted of two chemists and naturalists, and two scientific and practical apothecaries, who were paid by the government and played no part in the instruction of the candidate. The exam took place over the course of eight days, which included translating randomly selected parts of the *Prussian Pharmacopoeia* from the Latin, conducting a complete chemical analysis of two random substances, identifying samples of plants, drugs, and several chemical preparations while conducting purity tests. All the while, the examiners observed and recorded the successes and failures of the candidate. If they approved of the candidate's performance by a majority vote, they recommended the candidate for the public oral examination in subjects including chemistry, natural history, and medical law. If the candidate succeeded in passing the oral examination, he was recommended to receive a license to practice.[2] The rigorous training described above was for the first class apothecaries, who could practice anywhere in Prussia. There was also a less rigorous apprenticeship program for second-class apothecaries who, if licensed, could only practice in the local small towns and the countryside where they took their examinations.

The Roots of Modern Pharmacy Education in France

Similarly, the French government passed the Law of Germinal an XI in 1803, which established the modern system of French pharmacy education by founding pharmacy schools in Paris, Montpellier, and Strasbourg, which later merged with local universities in the 1840s. Later in the century, a pharmacy school was founded in Nancy and these four pharmacy schools remain the centers for pharmacy education in France.

In 1803, the pharmacy curriculum called for four courses including botany, the natural history of medicines (*materia medica*), chemistry, and practical pharmacy. In order to become an apothecary, the candidate must have served an apprenticeship of eight years or have served an apprenticeship of six years with credentials demonstrating attendance of three courses of lectures in a pharmacy school. For licensure, the candidate had to show certificates from the school attended as well as from the pharmaceutists he served under, as well as a birth certificate showing he was at least 25 years of age. If the documentation was found to be satisfactory, then a series of three public examinations was scheduled for a period of not less than three months. The candidate was examined on botany, the natural history of drugs, chemistry, and practical pharmacy. A favorable two-thirds of votes from the examiners, was required to receive a license. There was also a second-class apothecary program established by the 1803 law, which established preparatory schools to prepare apprentices for local examinations. These apothecaries had to limit their practices to small towns and rural areas.[3]

Early British Pharmacy Education

Not influenced by Continental traditions, British pharmacy education took a different approach from their Prussian and French counterparts. Instead, the first regulation to guide British pharmacy was the King James I's charter which created the Society of Apothecaries of London in 1606. The charter still recognized grocers as part of this society until Sir Francis Bacon interceded with King James I on the behalf of apothecaries, which resulted in their being recognized as "Master, Wardens and Society of the Art and Mystery of the Apothecaries of the City of London" on December 6, 1617. In addition to granting official recognition to apothecaries as a separate entity, the new charter called for aspiring apothecaries to serve seven-year terms of apprenticeship before they were eligible to take the examination. The examination tested the candidate's knowledge about the compounding of various medicines.[4]

While botanic courses were offered as early as 1627, they were followed by lectures in *materia medica* in 1753. These courses formed the basis of the first standard curriculum developed by the Society of Apothecaries which appeared in 1827. Their curriculum dealt mostly with medical subjects including anatomy, physiology, and medicine. Along with the coursework, student apothecaries were obliged to serve five-year apprenticeships before becoming eligible to take their examinations for licensure.[5] Although the Apothecaries Act of 1815 reinforced official recognition of apothecaries as an integral part of the medical community, the law still allowed anyone who wanted to sell drugs and poisons to do so without a license until legislation restricted the sale of poisons in 1868. As long as the unlicensed did not claim to be druggists, apothecaries, or chemists, they were free to sell drugs to the public. Due to the prevailing laissez-faire atmosphere, law-makers left pharmacy education and licensure largely unregulated until 1918 when an official curriculum was developed for the nation.[6]

The establishment of the Pharmaceutical Society of Great Britain in 1841 proved to be a landmark event in the history of British pharmacy education. Dedicated to elevating the scientific education of its members, the Pharmaceutical Society founded its School of Pharmacy in London in 1842 and by 1844 was offering science courses with laboratory requirements. The flagship school attracted a talented faculty including Jonathan Pereira (1804–1853) and Theophilus Redwood (1806–1892) among others. During this era, a number of proprietary schools emerged whose sole purpose was to prepare candidates for their examinations and, by the twentieth century, these schools could not meet the needs of students who needed to understand the scientific advancements modern pharmacy required. The Pharmaceutical Society's School of Pharmacy merged with the University of London in 1949 and became the model for nine other pharmacy schools by offering a bachelor of pharmacy and one year of post-graduate training in pharmacy as the standard for admission to practice. Graduate degrees in various research fields related to pharmacy are offered at the Royal Institute of Chemistry under the auspices of the Pharmaceutical Society.[7]

Pharmacy Education in the United States

Pharmacy education in British Colonial America followed the rather chaotic and circuitous route of Britain. Complicated by the necessity of coping with the unique economic, geographic, and political challenges of settling a new land, trained physicians and apothecaries initially were quite scarce and thus the transition from apprenticeship to formal education took centuries to achieve. Similar to their European counterparts, early American pharmacy education was conducted by apprenticeship. Because of the scarcity of apothecaries, physicians would offer indenture contracts to aspiring apothecaries for terms of five to seven years. In exchange for instruction, room, and board the apprentice was required to work for the physician. These apprenticeships began around age 14 and often lasted until 21, a vestige of the old European guild system. Initially, the apprentice performed menial tasks such as running errands, cleaning utensils, and cleaning the shop until he learned how to compound medicines and tend to the counter. In addition to learning the practical aspects of the apothecary's art, young apprentices were taught basic reading, writing, and arithmetic. Upon completion of the indenture the apprentice was free to work with other physicians in "doctor's shops" or perhaps in an apothecary shop.[8]

There were a few early attempts to offer courses in pharmacy to apothecaries by Dr. Lewis Mottet in South Carolina in 1769, an apothecary named James Cutbush from Philadelphia in 1812, and Dr. James Mease from Philadelphia in 1816.[9] Despite their good intentions none of these attempts proved to be successful or sufficiently adequate to become self-sustaining. These first attempts at "rounding off" the apprenticeship experience with a few formal courses were not complete failures because in 1821, due to a strange turn of events in Philadelphia, the first college of pharmacy in the United States was founded.[10]

The Founding of the Philadelphia College of Pharmacy

Similar to its European counterparts, formalized American pharmacy education began in medical schools with physicians teaching physicians and, later, apothecaries about pharmacy, exemplified by the career of Dr. John Morgan at the Medical College of Philadelphia (see Chapter 9). With the rapid advances in chemistry and the growing complexity of compounding drugs, the need for specially trained apothecaries became apparent (see Chapter 11 for a full discussion of the founding and early days of the Philadelphia College of Pharmacy).

Early Pharmacy Handbooks and Texts

Unlike the Europeans, the Americans developed a dispensatory before they had a pharmacopoeia. Based on the *Edinburgh New Dispensatory*, Dr. John Redman Coxe edited and published his *American Dispensatory* in 1806, which went through nine editions. In 1810, Dr. James Thatcher of Boston based his *American New Dispensatory* on an American pharmacopoeia, the *Pharmacopoeia of the Massachusetts Medical Society*, which had been issued in 1808 and dealt with substances indigenous to North America.[11] The ubiquitous influence of the Philadelphia College of Pharmacy made itself felt with the publication of *The United States Dispensatory* in 1833, authored by two of its star professors Drs. George Wood and Franklin Bache. In 1847, Carl Friedrich Mohr, a professor of chemistry and pharmacy at the University of Bonn published *Lehrbuch der pharmazeutischen Technik*, which was translated and adapted into English by Theophilus Redwood, a professor of chemistry and pharmacy at the pharmacy school established by the Pharmaceutical Society of Great Britain. Redwood wrote the book for British pharmacists, *Practical Pharmacy*, founded on Mohr's *Manual* and published it in 1848. Similarly, William Procter, Jr. recognized a good thing when he saw it and adapted Mohr's and Redwood's work for American apothecaries and, in 1849 published *Practical Pharmacy: The Arrangements, Apparatus, and Manipulations of the Pharmaceutical Shop and Laboratory.*[12]

Edward Parrish

An enterprising young Philadelphia apothecary named Edward Parrish (1822–1872), bought a shop near the University of Pennsylvania Medical School in 1843. In 1849, he opened his School of Practical Pharmacy where he taught aspiring physicians how to compound medicines found in the pharmacopoeia. Parrish and his school transformed pharmacy education, because here was an instance of an apothecary teaching physicians about drugs, breaking centuries of the precedent of physicians teaching apothecaries about drugs. A product of his teaching experience was Parrish's textbook entitled *Introduction to Pharmacy*, which was published in 1856. The actual title provides the rationale for his book: *An Introduction to Practical Pharmacy: Designed as a Text-book for the Student, and as a Guide to the Physician and Pharmaceutist.* With many formulas and prescriptions, Parrish's book represented a landmark in the literature because it was the first entirely American textbook on pharmacy based on American sources. Parrish's book proved to be quite popular and went through several editions until another Philadelphia College of Pharmacy alumnus made his mark.[13]

Joseph Remington

Born into a Philadelphia Quaker family in 1847 from a line of apothecaries deeply rooted in the profession, Joseph Price Remington (1847–1918) became an icon in the history of pharmacy. After graduating from the Philadelphia College of Pharmacy in 1866, he worked in pharmaceutical manufacturing with Edward Squibb and others before joining the faculty at the Philadelphia College of Pharmacy, a position he held for 44 years. Similar to Parrish, Remington wrote his book based on his teaching experiences, but instead of focusing his efforts on teaching physicians about drug compounding, he focused on teaching pharmacists. In 1885, Remington published his iconic

Figure 12.1 Edward Parrish (1822–1872) wrote the first purely American pharmacy textbook, *Introduction to Practical Pharmacy* in 1855, was a founding and active member of the American Pharmaceutical Association, and became the first president of Swathmore College. His work in the APhA led apothecaries to be called "pharmacists" in 1867.

Source: Courtesy of the National Library of Medicine.

Practice of Pharmacy, which was later referred to as *Remington's Pharmaceutical Sciences* and now titled *Remington: The Science and Practice of Pharmacy* or simply known as "Remington's." Consistent with the nineteenth century's penchant for long subtitles Remington's subtitle was *A Treatise on the Modes of*

Making and Dispensing Officinal, Unofficinal, and Extemporaneous Preparations, with Descriptions of the Properties, Uses, and Doses. Intended as a Hand-Book for Pharmacists and Physicians and Text-Book for Students. Remington was known as a dynamic and devoted professor, who inspired a generation of American pharmacists through his teaching. To honor Remington's life and work, the APhA sponsors an annual award that is presented to a distinguished pharmacist, named the Remington Award.[14]

Another popular textbook that was used for several decades in Britain and the United States was James Attfield's *An Introduction to Pharmaceutical Chemistry*, which was published in 1867. Attfield was a professor of practical chemistry at the School of the Pharmaceutical Society of Britain. Subsequent editions of his book which lasted well into the twentieth century were renamed *Chemistry, General, Medical, and Pharmaceutical.* Similar textbooks were written by American authors including Frederick Hoffman in 1873 and John Uri Lloyd in 1881.[15]

Pharmacy Education Evolves

Although the Philadelphia College of Pharmacy established a new paradigm for pharmacy education in America, most practitioners still believed that pharmacy was best learned through apprenticeships and the early proprietary schools of pharmacy struggled. Most apothecaries continued to view these schools as "finishing schools" to "round off" the education of the apprentice. Unlike medical schools in which classes were held during the day and required full-time attendance, pharmacy schools continued to offer part-time evening programs into the early 1900s. Typically, in its first 20 years of operation the Philadelphia College of Pharmacy graduated an average of 21 students per year.[16] Philadelphia was followed by the founding of the Massachusetts College of Pharmacy in Boston in 1823. Massachusetts offered occasional evening lectures and discussion sessions until 1867, when it began to offer a regular sustained course of evening instruction. While Philadelphia had consistent enrollment, it only began to offer laboratory instruction in 1870. By 1865, there were six schools offering degrees in pharmacy and the Tulane Medical School that offered a course in pharmacy. On the eve of the Civil War there were about 500 apothecaries who had earned degrees from American schools.[17] Still, by 1900, there were more than 60 schools offering pharmacy programs. In 1900, there were approximately 46,200 practicing pharmacists in the United States. By 1920, there were 66,200.[18]

Encouraged by the examples set by the Philadelphia College of Pharmacy as well as the founding of the APhA in 1852, several local pharmaceutical and medical associations emerged expressing an interest in advancing pharmacy education. Several of these pharmaceutical associations played key roles in establishing colleges of pharmacy in Boston (1823), New York (1829),

Baltimore (1840), Cincinnati (1850), Chicago (1859), and St. Louis (1864). Similar to the Philadelphia College of Pharmacy, each of these new institutions was founded by dynamic visionaries who viewed pharmacy education as vital to recognition for pharmacy as a profession in the new nation and also toward promoting its long-term survival. Many of these visionaries, such as Daniel B. Smith, William Procter, Jr., Edward Parrish, and others, were well aware of how far behind American pharmacists were at this time compared to their European counterparts and viewed pharmacy education and the building of professional pharmacy organizations as vehicles to achieving the goal of making American pharmacy the envy of the world within a few generations. Unfortunately, these visionaries confronted a long-standing opposition toward formalizing pharmacy education in the United States; the aspiring apothecaries and their preceptors themselves. Amid a laissez-faire atmosphere that permeated every aspect of American life, the average apothecary at this time learned his trade through apprenticeship and saw little need for formal education. In fact, prior to the 1870s there was virtually no regulation over who could open a shop and sell drugs. The numbers speak for themselves. Robert Buerki observed that "of the 11,031 apothecaries and druggists in the United States in 1860, only 514 had graduated from a pharmacy course, most from Philadelphia."[19] Moreover, Michael Flannery and Dennis Worthen noted that by 1900 about 12 percent of American pharmacists had graduated from a formal pharmacy program.[21]

At the 1866 annual meeting of the APhA, Edward Parrish from the Philadelphia College of Pharmacy proposed that all apothecaries, druggists, and pharmaceutists be known as "pharmacists." In 1867, the members of the APhA adopted Parrish's proposal and from that point forward the term "pharmacist" became official.[20]

Pharmacy Education after the Civil War

After the Civil War the nation was rebuilding and entering the age of industrialization in earnest which required an increasingly skilled and educated labor force, and this held especially true for pharmacy. While the Civil War had created massive destruction and displacement it also transformed the way pharmacy would be practiced with the introduction of mass-produced drugs by pharmaceutical manufacturers led by companies such as Squibb and Lilly. This was also the advent of the era of germ theory, which would give birth to the modern health care system that would place pharmacists squarely on the frontlines of health care. In addition to the established proprietary pharmacy schools, state-supported pharmacy schools began to appear that were part of a state university system. The first state-supported school was the Medical College of the State of South Carolina which opened in 1867.[22] During the

Civil War, a landmark piece of legislation called the Morrill Land Grant Act was passed by the Congress in 1862. This law opened up access to higher education by taking some of the proceeds earned by the government from western land sales and using them to establish state universities that would be open to all who qualified, chiefly in the Midwestern states.[23]

Albert Benjamin Prescott and the University of Michigan

The first of the Midwestern universities to offer a pharmacy program was the University of Michigan in 1868 under the direction of one of pharmacy's most gifted visionaries, Dr. Albert Benjamin Prescott (1832–1905). A physician and chemist, Prescott was charged with developing a full-time academic pharmacy program that would feature two years of graded laboratory work in the basic sciences and pharmacy. The new program featured the new fields of science that emerged during this time, namely organic

Figure 12.2 Albert Benjamin Prescott (1832–1905) transformed American pharmacy education by creating a full-time university-based program of study leading to a pharmacy degree in 1868 at the University of Michigan. Within 30 years, every pharmacy program in the country followed the model he established.

Source: Painting by Robert Thom, reproduced courtesy of the American Pharmacists Association Foundation.

chemistry, bacteriology, and pharmacology.[24] Prescott's program became a source of great controversy at the APhA's Annual Meeting in St. Louis in 1871, because it rejected the idea of student apprenticeships. Bowing to pressure from the older colleges of pharmacy, Prescott was allowed to attend as member, but not as delegate due to the fact that his program "was not controlled by pharmacists, nor an institution of learning, which by its rules and requirements insures its graduates the proper practical training."[25] Ostracized by pharmacy educators for nearly 28 years, Prescott was vindicated when he served as President of the APhA in 1899–1900 and was elected as the first president of the Conference of Pharmaceutical Faculties (now the American Association of Colleges of Pharmacy). Prescott's model was adopted by Frederick B. Power at the University of Wisconsin in 1883 and by six other state university programs by 1900, although some of these programs, including Wisconsin, required some apprenticeship time for their students. Under Power's leadership, the University of Wisconsin raised the admission requirement to the pharmacy program in 1884 to either a high school diploma or previous college attendance.[26] Balancing the needs of a strong academic background with the need for practical experience continues to be a strong subject of debate among pharmacy educators.

Albert Benjamin Prescott (1832–1905)

Prescott was born in Hastings, a small town in upstate New York. He was descended from a family that counted a military commander at the Battle of Bunker Hill and a prominent historian. Prescott injured his knee as a youth, a lifelong injury that compelled him to walk with a cane. After graduating from medical school at the University of Michigan in 1864, Prescott served as an army surgeon. After his service he returned to the University of Michigan where he taught chemistry. Prescott led the effort to establish a school of pharmacy at the University of Michigan and was named as its first dean. Prescott envisioned a two-year full-time course of study with lots of laboratory work, but no apprenticeship requirement. He endured ostracism by the more established schools for the lack of an apprenticeship requirement, but Prescott remained undaunted and by the turn of the twentieth century the pharmacy profession embraced Prescott's vision for placing pharmaceutical education on a par with other university programs and degrees. When he died in Ann Arbor on February 15, 1905, Prescott left behind a wife and an adopted son, and had transformed American pharmacy education.[27]

Edward Kremers and the University of Wisconsin

Several of the state universities and land grant colleges took the lead in advancing pharmacy education by raising admission standards to their programs and lengthening the course of their academic programs. In 1892, the University of Wisconsin took bold steps under its new director Edward Kremers (1865–1941), who for the remainder of his career became one of the greatest educational reformers in the history of pharmacy education. Kremers improved the two-year program by expanding it to three full terms of study for each year and strengthening the thesis requirement. Facing opposition from other departments at his own university as well as criticism from other pharmacy programs, Kremers instituted the nation's first four-year program in pharmacy as an option for students in 1892. The goal of the four-year program was to place the pharmacy program on a par with other collegiate degrees. The only other state university programs to adopt the four-year program for pharmacy were Ohio State, Georgia, Minnesota, and Nebraska.[28] In 1893, the University of Wisconsin awarded its first master's degree in pharmacy followed in 1902 by the first doctorate in pharmaceutical chemistry. The Ph.D. in pharmaceutical chemistry was a degree designed to prepare students to teach future faculty members and scientists. In 1917, the University of Wisconsin awarded its first Ph.D. in pharmaceutics.[29]

Edward Kremers (1865–1941)

Edward Kremers was born to Prussian immigrant parents in Milwaukee on February 25, 1865. He went to the Philadelphia College of Pharmacy but after a year returned to the University of Wisconsin where he earned a bachelor of science degree and later a doctorate in 1890. Having studied in Germany, Kremers became acquainted with a model of pharmaceutical education that he wanted to institute at the University of Wisconsin. Kremers envisioned a model for both undergraduate and graduate education in pharmacy and had the courage to bring this vision to fruition. In addition to his pioneering efforts in reforming pharmacy education he had a deep, abiding appreciation for the importance of the history of pharmacy. His textbook, *Kremers and Urdang's History of Pharmacy*, remains a classic in the field. Kremers retired in 1935 and died in 1941, a tireless champion for the advancement of pharmacy education. In 1941 the American Institute of the History of Pharmacy was founded, in part, as a tribute to Kremers.[30]

Alternative Pharmacy Education through Self-Study

With the establishment of state pharmaceutical associations across the nation beginning with the Maine Pharmaceutical Association in 1867, several of these associations called for legislation to regulate the practice of pharmacy and, more importantly, to define who could practice. While establishing state examinations for licensure, most states had "grandfather" clauses that awarded licensure to pharmacists who were already practicing following a precedent that had already been established for physicians. Facing state licensure requirements, many aspiring pharmacists searched for alternatives to attending pharmacy schools. In addition to apprenticeship and self-study, correspondence schools, cramming schools, and self-paced reading courses promised to fill this void. The first correspondence course was offered by the "National Institute of Pharmacy," which was operated by the publisher of *The Western Druggist* in 1885. The course was developed by Carl Svante Hallberg (1856–1910), a professor at the Chicago College of Pharmacy, and consisted of 24 lectures that covered *materia medica*, chemistry, and pharmacy that could be completed within a year. Building on Hallberg's program, another course that spanned two years of correspondence lectures and examinations was developed by the journal *Pharmaceutical Era* in 1897. "The Era Course of Pharmacy" was pioneered by James H. Beal (1861–1945) who was the dean of the Scio College Department of Pharmacy in Ohio. This course featured lectures by some of the leading professors of American pharmacy during that time, including John Uri Lloyd and Edward Kremers. There were also more dubious programs exemplified by "Lincoln-Jefferson University" out of Chicago that offered ten lessons that would be mailed to the student after a cash payment was received. This program promised attractive diplomas that would not reveal that the degree was based on taking a correspondence course. Surprisingly, this program survived until 1926.[31]

An alternative to the correspondence courses was the self-study book. One of the most popular and enduring of these books was published by Francis E. Stewart in 1886 entitled, *A Compend of Pharmacy*. Another popular book was *A Course of Home Study for Pharmacists*, which was written in 1891 by Oscar Oldberg (1846–1913), who was the dean of the Illinois College of Pharmacy at Northwestern University. As a result of the success of the book, Oldberg helped establish a series of university extension courses mostly in the basic sciences. Influenced by British efforts in this area, several schools over time converted these "college extension" courses into continuing education programs that nearly every pharmacy school offers today to help pharmacists remain current and proficient.[32]

Pharmacy Education in the Twentieth Century

At the dawn of the twentieth century, there were several seemingly unrelated events that gradually transformed pharmacy education: the forming of the American Conference of Pharmaceutical Faculties in 1900, the founding of the National Association of Boards of Pharmacy (NABP) in 1904, the New York state law passed in 1905 that required college graduation for pharmacy licensure, and Wilbur L. Scoville's survey of American pharmacy schools and the host of different requirements they set for the variety of degrees they offered in pharmacy. The confluence of these events challenged the AphA, the newly founded American Conference of Pharmaceutical Faculties (today's AACP), and the NABP to work together to find common cause in agreeing on a standard curriculum and a standard degree offering for pharmacists. This would prove to be quite challenging for all of the parties involved.

In 1905, Wilbur L. Scoville of the Massachusetts College of Pharmacy, conducted a survey of 78 American pharmacy schools and found a great disparity in the length of their curricula as well as their degree offerings. For example, in the case of the Ph.G. (Graduate in Pharmacy) degree the range was anywhere from three months to three years! The Doctor of Pharmacy (Pharm.D.) degree ranged from two to four years. The Ph.C. or Pharmaceutical Chemist degree ranged from one to five years. The Bachelor of Science or B.S. in Pharmacy degree (B.S. Pharm.) ranged from two to four years. This disarray of curricular requirements begged for reform. Perhaps most astonishing was that, as late as 1913, most of these schools only required one year of high school as a requirement for admission to their programs.[33]

To address this crisis in pharmacy education, pharmacy leaders turned to Abraham Flexner who had recently completed his landmark study on medical education, which was funded by the Carnegie Foundation. In 1915, pharmacy leaders asked Flexner to conduct a similar comprehensive study of pharmacy education and were shocked at his refusal on the grounds that he did not consider pharmacy to be a profession. Flexner was not alone in this belief. Since 1894, the American military would not grant military pharmacists commissions as officers until federal legislation proposed by Congressman and pharmacist Carl Thomas Durham (1892–1974) was passed in 1943, establishing a Pharmacy Corps for the army.[34]

Undaunted, in 1923 pharmacy leaders recruited W.W. Charters, a professor at the University of Pittsburgh, to conduct a study. Working from a grant funded by the Commonwealth Fund, Charters surveyed what pharmacists actually did on the job and the knowledge they needed to do their job. Typical of medical education in general at the time, little thought was given to what patients needed from their pharmacists. During the course of his study, Charters observed that the pharmacy profession had in many cases allowed itself to be overwhelmed by business concerns at the expense of patient care. He was also disturbed to find many pharmacists engaged in "counter-prescribing," i.e. pharmacists dispensing medicines to patients directly without a physician's prescription, which he viewed as dangerous given the lack of formal education. With the assistance of A.B. Lemon, Leon M. Monell, and Dr. Robert P. Fischelis, Charters published a book entitled *Basic Material for a Pharmaceutical Curriculum*, which outlined what was needed for a sound baccalaureate degree in pharmacy. In 1927, Charters published the study and recommended increased educational requirements that led to

Table 12.1 Pharmacy Degrees

1907	Ph.G.	2 years
1925	Ph.C.	3 years
1932	B.S.	4 years
1950	B.S.	5 years
1989	Pharm.D.	6 years
1st Pharm.D. programs 1950 at USC and 1954 at UCSF.		

AACP's adoption of a four-year baccalaureate degree that became effective for all AACP member institutions in 1932. Charters also concluded that pharmacy was a profession.

State Licensure

While there were calls for a national policy that required candidates who sat for state licensure examinations to hold degrees from a pharmacy school as early as 1891, nothing tangible was accomplished to make this happen. What could not be agreed upon at the national level eventually was achieved at the state level. The first step in this direction occurred when the New York State Legislature passed a law requiring candidates for licensure to be 21 years old, have four years of practical pharmacy experience, and pass a comprehensive examination. The law also charged the state's pharmacy schools with teaching a two-year course of study that met "a proper pharmacy standard."[35] The next year the Pennsylvania Legislature passed a similar law requiring four years of practical pharmacy experience and graduation from a pharmacy school.

These state laws, especially New York's, acted as a catalyst to get the pharmacy schools to work toward standardizing their curricula and their degree offerings. The NABP and the ACPF, came together to organize a National Syllabus Committee at the request of representatives from New York.[36] The schools that were members of the ACPF adopted the two-year curriculum in 1907 as the requirement for the Ph.G. degree. In 1910, the National Syllabus Committee published its findings and called it, "The Pharmaceutical Syllabus" which established a fairly standard two-year pharmacy curriculum recommending 1,200 hours of instruction as a minimum. According to Robert Buerki, "By 1913, 62 of the 83 schools and colleges of pharmacy in the United States had formally adopted the Syllabus; a second edition was published in 1913, a third in 1922, and a fourth in 1932."[37]

When ACPF became the American Association of Colleges of Pharmacy in 1925, the organization instituted high school graduation as an entrance requirement to its member institutions. AACP also adopted a three-year curriculum for its members in 1925 and approved a four-year program, which was to become effective in 1932. The NABP hoped to conduct a broader survey of pharmacy, but the Great Depression and World War II delayed these plans. Despite all of these achievements, the demands of World War II helped to sweep away all of this progress as pharmacy schools struggled mightily for enrollment because the men were off to war and few women enrolled. In order to help the war effort, some pharmacy programs offered year-round accelerated programs. For example, the St. Louis College of Pharmacy offered its four-year bachelor's degree in two years and eight months by having students attend three continuous 16-week semesters per year. By 1946, the Pharmaceutical Syllabus became a dead letter as its sponsoring organizations abandoned it.

The Pharmaceutical Survey 1946

After World War II, as the nation tried to return to peace, pharmacy leaders wanted a self-study of the profession and on April 15, 1946 called upon Edward C. Elliott, president emeritus of Purdue University and a non-pharmacist, to conduct a study. Elliott's charge was to conduct a wide-ranging study of pharmacy in all of its aspects that came to be known as the Pharmaceutical Survey. Similar to Charters, Elliott focused on what pharmacists did and needed instead of what patients needed. *The General Report of the Pharmaceutical Survey 1946–1949* agreed with the Charters Commission Report that pharmacy was indeed a profession, but warned that if significant improvements were not made, especially in strengthening the curriculum, pharmacists risked losing this professional status. Specifically, the report argued that the current four-year curriculum was inadequate and recommended it be expanded to a six-year doctoral degree that would provide sufficient time to provide students with a sound background in the basic sciences as well as the general education associated with a university degree. As expected, these recommendations were met differently by different constituencies in the pharmacy community. Educators, hospital pharmacists, and most pharmacy organizational leaders supported the recommendations, while NARD and community pharmacists remained skeptical about them. The debate about broadening the liberal arts education of pharmacists to gain professional and public respect continues today even though AACP adopted the six-year doctor of pharmacy degree in 1989. Ultimately, another compromise was reached and all accredited pharmacy schools implemented the doctor of pharmacy degree with the entering class of 2000. AACP brokered a compromise and endorsed a five-year curriculum in 1954, which would take effect in 1960. In 1950, the University of Southern California started the first six-year program in the nation followed by the University of California–San Francisco in 1954. The Ohio State University adopted the five-year degree in 1948.[38] Many historians view the Pharmaceutical Survey as a visionary study with cogent recommendations for the advancement of pharmacy that took far too long for the pharmacy community to implement.

The Millis Commission

In 1972, the President of the American Association of Colleges of Pharmacy (AACP) Arthur Schwarting, the Dean of the College of Pharmacy at the University of Connecticut, called for a commission to study the future role of pharmacy amid the rapidly changing American health care system. AACP called upon John S. Millis, a former chancellor of Case Western Reserve University and a veteran leader of several other health care related studies, to lead the commission. The Study Commission on Pharmacy's composition was broad-based including academic, community, and hospital pharmacists as well as physicians, corporate executives, academics, and a nurse. Unlike previous commissions, Millis' group set out to study pharmacy from an external, rather than internal perspective. The Millis Commission wanted to know what patients and other health care professionals wanted pharmacists to be able to do, rather than studying what pharmacists were already doing for patients.

After meeting for two years with many key constituencies in the health care system, the Millis Commission released its report in December 1975, entitled *Pharmacists for the Future*. The report called upon the pharmacy profession to end the educational practices that isolated pharmacy students from dealing directly with patients and other health care professionals. While medical and nursing education had significant patient contact components, pharmacy education did not. The report also called for pharmacists to abandon their traditional focus on drug products and shift it to interacting with patients in terms of educating them about how to get the most from their drug therapies. The report clearly endorsed the practice of clinical pharmacy as the profession's future.[39]

While the Millis Commission's recommendations did not provide a detailed blueprint for reforming pharmacy education, it inspired members of APhA and AACP to provide such details that resulted in reformatting the pharmacy curriculum. *The National Study of the Practice of Pharmacy* in 1979, sponsored by APhA and AACP, defined and established standards of practice that provided important guidance to curriculum committees and accreditation bodies alike. In 1984, the APhA Task Force on Pharmacy Education in its *Final Report* agreed upon the minimum competencies expected of an entry-level pharmacist and called for the implementation of a six-year doctor of pharmacy degree as the entry-level degree for all pharmacists.

Pharmaceutical Care

These two reports gained endorsement from an invitational conference held at Hilton Head Island, South Carolina in 1985. The conference's title "Directions for Clinical Practice in Pharmacy" called for pharmacy to become a true clinical profession. At the AACP annual conference held in Charleston, South Carolina, in 1987, C. Douglas Hepler introduced the concept of "pharmaceutical care." In 1989, at the second "Pharmacy in the 21st Century Conference" held in Williamsburg, Virginia, Hepler teamed with Linda Strand to define pharmaceutical care as "the responsible provision of drug therapy for the purpose of achieving definite outcomes that improve a patient's quality of life."[40] Moreover, pharmaceutical care was a concept that could be applied to all sites where pharmacy was practiced.

The concept of pharmaceutical care gained acceptance and necessitated changes in the pharmacy curriculum to refocus the pharmacist's attention from the drug product to the patient and more importantly called upon the pharmacist to assume responsibility for the patient's health outcomes. The concept of pharmaceutical care reframed the terms of the decades old debate about the B.S. or the Pharm.D. as the ideal entry-level degree for practicing modern pharmacy. In July 1989, AACP President William Miller, established the Commission to Implement Change in Pharmaceutical Education. The Commission's goal was to recommend a pharmacy curriculum that would provide students with the knowledge, skills, and attitudes to deliver sound pharmaceutical care. Two months later, the American Council on Pharmaceutical Education (ACPE) released new accreditation standards that would recognize the Pharm.D. as the sole entry-level degree that would take effect with the entering class of 2000. This was a landmark transformation of pharmacy education and a fulfillment of the recommendations issued by the Elliott Commission back in 1954.[41]

During the early 1990s, the AACP Commission issued two White Papers that defined the mission of pharmaceutical education and dealt with curricular content and outcomes. The Commission's White Paper "Entry-level Education in Pharmacy: A Commitment to Change" called upon educators and the pharmacy community to embrace the spirit of change and to support the outcomes associated with the delivery of pharmaceutical care. In July 1992, the AACP House of Delegates officially endorsed the doctorate as the single entry pharmacy degree. While the Pharm.D. and the concept of pharmaceutical care enjoyed broad support, a challenge appeared that caused considerable controversy and further debate. Dealing with what appeared at the time to be ominous manpower projections for the pharmacy profession, the Pew Health Professions Commission issued its report in 1995. The report called for a 20–25 percent reduction in the number of pharmacy schools, which at the time meant closing anywhere from 15 to 18 schools, presumably those that only offered the Bachelor of Science in Pharmacy degrees. The report estimated that there would be an excess of nearly 40,000 pharmacists due to the automation and centralization of dispensing functions. The report praised the concept of pharmaceutical care's commitment to patient-centered care and viewed it as a model for other health professions to adopt. What the Pew Commission report did not anticipate was the retirement of the baby-boomer generation and the Medicare drug legislation in 2003 that would require more pharmacists, not fewer. Contrary to the Pew Commission's recommendations calling for substantial reductions in the number of pharmacy schools, there has been a sharp increase, with over 130 as of 2014, representing an increase of more than 60 schools since 1995.[42]

Medication Therapy Management

Inspired by the concept of pharmaceutical care in the 1990s the scope of clinical pharmacy expanded, especially in hospital settings. Medication and dosage errors became a source of media focus as the result of several studies conducted by the Dana-Farber Clinic in Boston, Harvard Medical School, Ken Baker at Auburn University, and others. By assigning clinical pharmacists to patient care teams with physicians and nurses, evidence mounted that sound pharmaceutical care practices could dramatically decrease medication errors in hospitals. A 1996 study conducted by Schumock reported that the median benefit-to-cost ratio of clinical pharmacy was 4.1 : 1; meaning for every dollar spent, a patient benefit of $4.10 was realized in better health outcomes.[43] More recent studies have demonstrated similar beneficial outcomes in community pharmacy and ambulatory care settings.

When Congress passed the Medicare Prescription Drug, Improvement and Modernization Act in 2003 for senior citizens, the concept of medication therapy management (MTM) was born. With 11 pharmacy organizations working together the following definition of MTM was developed: "Medication therapy management is a distinct service or group of services that optimize therapeutic outcomes for individual patients. Medication therapy management services are independent of, but can occur in conjunction with, the provision of medication product."[44] By 2007, the American Medical Association included codes that would compensate pharmacists for cognitive services (medication advice) in its *Current Procedural Terminology* that created a mechanism for MTM to become a reality.

The implications of pharmaceutical care and MTM hold enormous challenges for pharmacy schools as clinical pharmacy takes shape. With newer and more complex drug therapies coming to market in recent years several pharmacy schools have responded to this challenge by expanding their curricula beyond the traditional six years. The majority of pharmacy students entering pharmacy school today have already earned a bachelor's degree and those schools that admit students from high school have already expanded, or are in the process of expanding, their programs to seven or eight years.

Chapter Summary

Pharmacy education in the United States has come a long way from the old British medieval apprenticeship system. From the founding of the Philadelphia

College of Pharmacy in 1821, to the founding of the APhA, AACP, and ACPE, to the introduction of the concepts of pharmaceutical care and MTM, the apothecary, pharmaceutist, druggist, and pharmacist has adapted to these innovations in the service of their patients. The shift from the focus on the drug product to patient-centered care has challenged and transformed pharmacy education in America.

Key Terms

Pharmaceutical Society of Great Britain

John Morgan

John Redman Coxe

Philadelphia College of Pharmacy

Christopher Marshall

Daniel B. Smith

Ph.G.

The American Journal of Pharmacy

William Procter, Jr.

The United States Dispensatory

Carl Friedrich Mohr

Theophilus Redwood

Edward Parrish

Joseph Price Remington

Morrill Land Grant Act

Albert Benjamin Prescott

Frederick Power

Edward Kremers

The Western Druggist

Wilbur L. Scoville

Carl Thomas Durham

W.W. Charters

The National Boards of Pharmacy (NABP)

The American Association of Colleges of Pharmacy (AACP)

The Pharmaceutical Survey of 1946–1949

The Millis Commission

Hilton Head Conference 1985

Pharmaceutical Care

C. Douglas Hepler

Linda Strand

American Council of Pharmacy Education (ACPE)

Pew Health Professions Commission 1995

Medication Therapy Management (MTM)

Medicare

Prescription Drug, Improvement and Modernization Act 2003

Chapter in Review

1 Explain the European influence on the origin of American Pharmaceutical Education.

2 Describe a typical apothecary's education during the Colonial Era.

3 Trace the origin of the Philadelphia College of Pharmacy and its historical significance to American pharmaceutical education.

4 Describe how early American colleges of pharmacy operated in terms of what was taught and the textbooks students read.

5 Explain the significance of the Land Grant Universities and the reforms of Albert Prescott and Edward Kremers on American pharmaceutical education.

6   Describe the influence of the work of W.W. Charters, Edward Elliot, and John Millis and their commissions on the development of pharmaceutical education in the United States.

7   Explain the roles that AACP, ACPE, and NABP played and continue to play in pharmaceutical education.

8   Describe the circumstances under which the concept of "pharmaceutical care" originated and how it evolved.

9   Explain the significance of medication therapy management.

Notes

1   David L. Cowen and William H. Helfand, *Pharmacy: An Illustrated History* (New York: Harry Abrams, 1990), 145.

2   Charles Lawall, *The Curious Lore of Drugs and Medicines: Four Thousand Years of Pharmacy* (Garden City, NY: Garden City Publishing, 1927), 447–448. From President Daniel B. Smith's commencement address to the Philadelphia College of Pharmacy's class of 1929.

3   Ibid., 442–443.

4   George Bender, *Great Moments in Pharmacy* (Detroit, MI: Northwood Institute Press, 1966), 69–70.

5   Glenn Sonnedecker, comp., *Kremers and Urdang's History of Pharmacy* (4th ed., Madison, WI: AIHP, 1976), 115.

6   Ibid., 104–108.

7   Ibid., 117. David L. Cowen and William H. Helfand, *Pharmacy*, 146.

8   Gregory Higby, "A Brief Look at American Pharmaceutical Education Before 1900." In Robert A. Buerki, *In Search of Excellence: The First Century of the American Association of Colleges of Pharmacy* (Alexandria, VA: AACP, 1999), Fall Supplement 1999, Vol. 63, 1.

9   Glenn Sonnedecker, *Kremers and Urdang's History of Pharmacy*, 227.

10 Ibid.

11 Glenn Sonnedecker, *Kremers and Urdang's History of Pharmacy*, 278.

12 Ibid., 282–283.

13 Dennis Worthen, *Heroes of Pharmacy* (Washington, DC: APhA, 2008), 154.

14 Ibid., 175. Glenn Sonnedecker, *Kremers and Urdang's History of Pharmacy*, 284–285.

15 Ibid. (Sonnedecker), 286–287.

16 Ibid., 229.

17 Ibid., 244.

18 Jane Mobley, *Prescription for Success: The Chain Drug Story* (Kansas City: The Lowell Press, 1990), 4.

19 Robert Buerki, "Pharmaceutical Education, 1852–1902." In *American Pharmacy: A Collection of Historical Essays*, edited by Gregory Higby (Madison, WI: AIHP, 2005), 38.

20 Dennis Worthen, *Heroes of Pharmacy*, 155.

21 Michael Flannery and Dennis Worthen, *Pharmaceutical Education in the Queen City: 150 Years of Service 1850–2000* (New York: Pharmaceutical Products Press, Inc., 2001), 8.

22 Glenn Sonnedecker, *Kremers and Urdang's History of Pharmacy*, 232.

23 Ibid., 232. Robert Buerki, "American Pharmaceutical Education, 1852–1902," 38.

24 John Prascandola and John Swann, "Development of Pharmacology in American Schools of Pharmacy," *Pharmacy in History*, Vol. 25 (1983), No. 3, 98.

25 George Bender, *Great Moments in Pharmacy*, 141.

26 Dennis Worthen, *Heroes of Pharmacy*, 114.

27 Ibid., 161–165.

28 Glenn Sonnedecker, *Kremers and Urdang's History of Pharmacy*, 240.

29 Dennis Worthen, *Heroes of Pharmacy*, 115.

30 Ibid., 113–119.

31 Glenn Sonndecker, *Kremers And Urdang's History of Pharmacy*, 245.

32 Ibid., 246.

33 Robert M. Elenbaas and Dennis Worthen, *Clinical Pharmacy in the United States: Transformation of a Profession* (Lenexa, KS: ACCP, 2009), 131.

34 Dennis Worthen, *Heroes of Pharmacy*, 83.

35 Robert Buerki, "Pharmaceutical Education, 1902–1952," 44.

36 Robert Elenbaas and Dennis Worthen, *Clinical Pharmacy in the United States*, 13.

37 Robert Buerki, "Pharmaceutical Education, 1902–1952," 44.

38 Gregory Higby, "From Compounding to Caring: An Abridged History of American Pharmacy." In *Pharmaceutical Care*, edited by Calvin Knowlton and Richard Penna (2nd ed., Bethesda, MD: ASHP, 2003), 36. Robert Elenbaas and Dennis Worthen, *Clinical Pharmacy in the United States*, 133. Glenn Sonnedecker, *Kremers and Urdang's History of Pharmacy*, 253.

39 Ibid. (Sonnedecker), 254. Robert Buerki, "Pharmaceutical Education, 1952–2002," 51. Robert Elenbaas and Dennis Worthen, *Clinical Pharmacy in the United States*, 138–139.

40 C. Douglas Hepler, "Reflections on Clinical Pharmacy." In *Clinical Pharmacy in the United States*, edited by Robert Elenbaas and Dennis Worthen (Lenexa, KS: ACCP, 2009), 148.

41 Ibid., 157.

42 www.aacp.org (accessed September 20, 2014).

43 C. Douglas Hepler, "Reflections on Clinical Pharmacy," 163.

44 B.M. Bluml, "Definition of Medication Therapy Management: Development of Profession-wide Consensus," *Journal of the American Pharmacists Association*, Vol. 45 (2005), 566–572.

13 The Origin and Growth of the Pharmaceutical Industry

From *Terra Sigillata* to Viagra

How did prescription and nonprescription drugs become such an integral part of our lives?

Today it is difficult for us to imagine daily life without manufactured drugs. Vitamins, dietary supplements, over-the-counter drugs, and prescription drugs have become integral parts of our lives. By 2004, nearly half of all Americans were using at least one prescription drug daily.[1] Although we constitute less than 5 percent of the world's population the United States accounts for almost 50 percent of global sales of pharmaceuticals.[2] This chapter will trace the origins and evolution of the pharmaceutical manufacturing industry.

Terra Sigillata: The First Trademark Drug

The earliest known attempt at manufacturing drugs came with the ancient Greeks from the island of Lemnos who discovered a special type of clay they called *Terra Sigillata* or "sealed earth" *c.*500 BCE (see Chapter 4). The substance was used to treat dysentery, ulcers, gonorrhea, fever, and eye infections. Modern chemical analysis concluded that *Terra Sigillata* was composed of aluminum, silica, chalk, magnesium, and iron oxide all of which can be found in modern pharmacopoeias. *Terra Sigillata* was harvested once a year—dug from a hillside pit and taken to the village where it was washed and rolled into lozenges (pastilles) and impressed with a unique seal. This seal represented the first "trademarked" drug where patients were assured they were buying the genuine product. Thus, the proprietary drug business began. Luminaries including Hippocrates, Dioscorides, and Galen sang its praises as a wonder drug. Over time, similar greasy clay substances with all sorts of trademarks appeared, claiming unique cures for human ailments.[3]

Nearly all of the great physicians and apothecaries mentioned in the previous chapters compounded their own remedies, but these were made by hand and all done on a relatively small scale compared to today's pharmaceutical industry. For example, Hippocrates offered patients teas made from willow bark (primitive aspirin), Galen produced cold cream, while later in medieval Arab medicine Avicenna gilded and silvered pills to make them easier to swallow. During the era of monastic medicine in medieval Europe some monasteries in Italy, France, and the German states produced distilled waters and perfumes.[4] Troches (lozenges) of treacle made from viper flesh were manufactured on a larger scale in Venice during the fifteenth and sixteenth centuries.[5] Similarly, at around 1600, the Duke of Schleswig began mass producing chemicals that he forced local apothecaries to purchase so that they could compound prescriptions for their patients. In the seventeenth century, the skilled German chemist John Rudolph Glauber (1604–1670) engaged in the manufacturing of sodium sulfate (Glauber's salt), ammonium sulfate, zinc chloride, and potassium chloride in Amsterdam.[6]

The first pharmaceutical manufacturing in England began with The Society of the Art and Mystery of the Apothecaries of the City of London. In 1623, the society established a cooperative of local apothecaries who agreed to produce galenicals and chemicals on a large scale. The cooperative incorporated in 1628 and by 1703 the society had become the exclusive suppliers of drugs to the English navy and later to the army. Similarly, in 1801 the East India Company entered into an exclusive contract to buy all of their drugs from the society, which lasted until the mid-nineteenth century.[7] The success of the society's pharmaceutical manufacturing venture inspired other nations to form similar cooperative efforts: the *Pharmacie Centrale de France* in 1852, the *Hageda*, in Germany in 1902, and city boards of health in the United States during the late nineteenth century.[8]

Ambrosius Gottfried Hanckwitz: "Godfrey"

In 1660, the "father of modern chemistry" Robert Boyle (1626–1691) invited a talented young German chemist, Ambrosius Gottfried Hanckwitz, to London. Together, they built a chemist's shop with a laboratory. Hanckwitz, who later became known as "Godfrey," transformed this shop into the world's leading producer of phosphorous, earning him international renown by 1700. According to a bill Godfrey wrote, "He continues to prepare faithfully all sorts of remedies, Chemical and Galenicals—Good Cordials as English Drops, Gaskoin's Powder, Essence of Viper and Essence for the Hair, together with Sundry Spirits and Arquebusade."[9]

Antoine Baume

As chemistry was coming into its golden age in France during the eighteenth century a young apothecary named Antoine Baume (1728–1804) transformed his shop into a manufacturing plant where he produced drugs with great technical skill. Born in humble circumstances, Baume became a Professor of Chemistry at the College of France when he was 25. A born innovator, he invented the hydrometer and "Baume's degrees" that bear his name. In 1757, he developed the first modern formula for ether and also improved the process of distillation. His most notable book, *Manuel de Chymie* (1763), was translated into several languages and became a classic in pharmaceutical chemistry. In 1775, Baume issued an 88-page price list that cataloged 2,400 items, 400 of which were chemical preparations that he made in his laboratories. In addition to his pharmaceutical plant, Baume established the first factory that produced ammonium chloride.[10]

The Era of Alkaloidal Chemistry

The turn of the nineteenth century ushered in the age of alkaloid and synthetic phytochemistry, which revolutionized pharmaceutical chemistry and soon would outstrip the individual apothecary's ability to compete with it. These advances combined with the mass production capacity of the Industrial Revolution eventually would shatter the age-old paradigm of the apothecary compounding individual prescriptions by hand. For example, Friedrich Sertuerner's discovery of morphine and Pierre-Joseph Pelletier's and Joseph-Bienaime Caventou's discovery of quinine were remarkable scientific triumphs that demanded great specialized skill and equipment to replicate on a large scale. Pelletier for one recognized this and in 1822, two years after his landmark discovery, opened an industrial plant to manufacture quinine. In the same year in Philadelphia, immigrants John Farr and Abraham Kunzi manufactured quinine in their laboratory. In 1823, another Philadelphia company, Rosengarten and Sons, followed suit.[11] They were followed by Johann Daniel Riedel who began manufacturing quinine in Berlin in 1827. In that same year another German pharmaceutist, Heinrich Emanuel Merck (1784–1855), working out of his family's shop in Darmstadt, published a treatise on the production of various alkaloids entitled, "Cabinet of Pharmaceutico-Chemical Novelties." Merck manufactured morphine, quinine, emetine, strychnine, and other plant extracts. In 1891, Merck and Company opened a pharmaceutical plant in the United States.[12]

Stanislas Limousin: Pharmacy Inventor

Similar to Antoine Baume a generation earlier Stanislas Limousin (1831–1887) was a force of nature who combined scientific knowledge with a spirit of inventiveness that transformed pharmacy. Limousin worked out of his Parisian apothecary shop to support his research. A prolific inventor, Limousin invented the medicine dropper, a system for coloring poisons to identify them, and wafer cachets. Wafer cachets were a welcome new dosage form before the mass production of hard gelatin capsules. The wafer cachet made it possible to deliver foul-tasting medicinal powders by encasing them

Table 13.1 Early Manufacturing Firms

1660	Robert Boyle (succeeded by Godfrey)—England
1741	Burgoyne, Burbidges & Company, Ltd.—Scotland
1752	Antoine Baume—France
1780	J.F. MacFarlan & Company—Scotland
1780	Savory and Moore, Ltd.—England
1790	Wright, Layman, & Umney, Ltd.—England
1795	Allen and Hanburys, Ltd.—England
1798	John Bell & Croyden—England
1807	Howard and Sons—England
1814	Daken Brothers, Ltd.—England
1814	J.D. Riedel—Germany
1817	Friedrich Wilhelm Sertuerner—Germany
1820	Joseph Pelletier and J.B. Caventou—France
1821	Thomas Morson and Son—England
1827	H.E. Merck—Germany
1831	Squire & Sons, Ltd.—England
1833	Stafford, Allen & Sons, Ltd.—England
1837	Hermann Trommsdorff—Germany
1851	Ernst Schering—Germany
1856	Friedrich Witte—Germany
1880	Stanislas Limousin—France
1880	Burroughs, Wellcome & Company—England[13]

between two hard shells made of rice starch. These inventions would have been sufficient for Limousin to have made the history books, but he also invented glass ampoules that could be sealed and sterilized to preserve solutions that could be delivered by hypodermic needles. He also developed the apparatus that allowed patients to inhale oxygen. Stanislas Limousin personified the qualities of a modest genius who would never quite understand the praise he richly deserved.[15]

> The first medicine to gain a patent was Epsom salts in 1698 in England.[14]

The Pharmaceutical Industry in the United States

During the British colonial era the American production of drugs was severely limited due to mercantilism. There were the rare exceptions of John Sears of Cape Cod who produced Glauber's salt and Epsom salt and the governor of Connecticut colony, John Winthrop, Jr., who made medicines for his colonists, but both remained cottage industries. George Urdang observed that the American pharmaceutical industry was born amid the American Revolution and transformed by the American Civil War and later by World Wars I and II.[16] If there is a silver lining to warfare, it is that it compels people to be creative and invent new ways of doing things to meet the immediate challenges posed by war. The first large-scale manufacturing of drugs in the United States came about as a result of the American Revolution. Early in the war Andrew Craigie, a Boston apothecary, was commissioned by the Continental Congress to serve the Continental Army as an apothecary and was later appointed as Apothecary General. With imports blockaded by the British and drug supplies dwindling, Craigie developed a plan to establish a laboratory and storehouse for the production of drugs for medicine chests at Carlisle, Pennsylvania. Craigie's plan was adopted by the Continental Congress and the first large-scale manufacturing of drugs in the United States began.[17]

Philadelphia: The Birthplace of the American Pharmaceutical Industry

Soon after the American Revolution, the city of Philadelphia, which played such a crucial role in the Revolution and the early years of the new Republic, became the birthplace of American pharmaceutical manufacturing and pharmacy in general. In 1786, Christopher Marshall, Jr. and his brother Charles Marshall began manufacturing ammonium chloride and Glauber's salt. This was followed by John Farr and Abraham Kunzi who produced quinine in 1822. Farr and Kunzi were joined by druggists Thomas Powers and William Weightman and the company was renamed Powers and Weightman. In 1823, Rosengarten and Sons produced quinine sulfate, followed later by morphine salts, piperine, strychnine, codeine, bismuth, and silver salts. During the age of consolidation, Powers and Weightman merged with Rosengarten and Sons and later were absorbed by Merck in 1927.[18] In 1826, another Philadelphia wholesale and retail druggist, Samuel Wetherill began his own company and manufactured tartaric acid, Rochelle salt, calomel, sulfuric ether, and sulfate of quinine. This was followed in 1838 by the Philadelphia retail druggist Robert Shoemaker who developed a manufacturing process for making plasters. He also manufactured glycerin. In 1841, George K. Smith, who owned a retail and wholesale pharmacy, and his accountant, Mahlon Kline, established a pharmaceutical factory. Smith and Kline later consolidated their operations with a perfume manufacturer Henry French and formed Smith, Kline, and French to form the basis of what would later became one of the world's leading pharmaceutical manufacturers.[19] Still another Philadelphia druggist, William R. Warner, invented the sugar-coated pill and began manufacturing them in 1856. Capitalizing on the discovery of the compressed pill or "tablet" in the 1860s by a Philadelphia wholesale druggist named Jacob Dunton, John Wyeth and his brother established John Wyeth and Brothers and began the large-scale production of tablets, which revolutionized how medications in the United States would be administered.[20] Many of these manufacturers would go on to supply the Union forces with pharmaceuticals during the Civil War. To meet demand, the federal government established pharmaceutical plants in Philadelphia and Long Island. Similarly, the Confederacy also opened plants in at least a dozen locations in several states.[21]

One pharmaceutical company that did not start in Philadelphia was the William S. Merrell Company. Merrell opened his pharmacy in Cincinnati in 1828 and later opened a factory that produced a line of products made from native ingredients that became popular with eclectic physicians. Similarly, a Baltimore pharmacist, A.P. Sharpe, and one of his clerks, Louis Dohme, produced a line of fluid extracts, syrups, and chemicals.[22]

Once again with the outbreak of the Civil War in 1861, war profoundly shaped the course of the American pharmaceutical industry. In 1858, a year after leaving the U.S. Navy, Edward R. Squibb (1819–1900), a former physician and pharmacist, established a pharmaceutical laboratory in Brooklyn, New York, to produce ether and chloroform. Armed with a contract to produce pharmaceuticals for the U.S. Army, Squibb and his new company were on the verge of success when a fire destroyed his laboratory and left him scarred for life. Undaunted, Squibb with the help of his physician friends who counted on the high quality of his products rebuilt his factory. By 1863, with the Civil War raging, Squibb's company produced a full line of pharmaceuticals for the Union Army and was able to increase his factory's capacity to meet the demand. A lifelong advocate of drug purity and safety, Squibb

Table 13.2 U.S. Pharmaceutical Companies and When They Were Founded

1855	Frederick Stearns & Company
1856	William R. Warner & Company
1858	E.R. Squibb & Sons
1860	Reed & Carnick
1860	Wyeth
1860	Sharp & Dohme
1863	Burroughs Bros. Manufacturing Company
1866	Parke, Davis, & Company
1870	Lloyd Brothers
1875	Chilcott Laboratories
1876	Eli Lilly & Company
1884	Lambert Company
1885	Norwich Eaton
1885	Johnson & Johnson
1886	Upjohn Company
1886	Armour Pharmaceuticals
1887	Bristol-Myers
1887	Merck (US)
1888	Abbott Laboratories
1888	G.D. Searle
1897	Becton Dickinson[23]

worked tirelessly to combat the proliferation of adulterated medicines. By the time of his death, Squibb and Sons was one of a very select group of pharmaceutical manufacturers which produced a full line of pharmaceutical products on a national scale.[24]

Another company that benefited greatly from the Civil War was Frederick Stearns and Company of Detroit. Stearns began his company in 1855 with a small laboratory in the back of his pharmacy. He specialized in fluid extracts. With the demand for pharmaceuticals prompted by the Civil War, Stearns expanded his company. He automated his factory using steam powered machines and became a model of pharmaceutical manufacturing. After the war, similar to Squibb, Stearns became concerned about the proliferation of shoddy patent medicines. His company introduced a line of nostrums that listed the names and quantities of the nostrum's ingredients long before the Pure Food and Drug Act of 1906. The idea worked and made its way to Europe.[25]

The post-war era ushered in a new age in pharmaceutical manufacturing and saw the rapid growth of many new companies. Parke, Davis, and Company got its start in the retail shop of a physician/pharmacist named Samuel Duffield. Duffield had studied chemistry in Germany. Interested in producing established and novel fluid extracts from botanicals, Duffield teamed up with H.C. Parke and George S. Davis and founded their company in Detroit in 1866. While several companies were producing fluid extracts they invented a process to standardize them. In the 1880s, Parke, Davis, and Company hired Albert Lyons, a chemist from the University of Michigan, who helped the company to develop 48 fluid extracts from plants their agents collected in North and South America. In 1902, the company was one of the first to establish its own research institute. Other companies, including Smith, Kline, and French (1893), Lilly (1911), and Upjohn (1913), established their own research divisions.[26]

After several business setbacks, a former Union artillery colonel, Eli Lilly (1838–1898), opened a small laboratory in Indianapolis in 1876. Lilly was a pharmacist who produced fluid extracts, elixirs, and syrups, mostly for sale to physicians. His son and successor, Josiah K. Lilly (1861–1948) also graduated from the Philadelphia College of Pharmacy and helped establish Eli Lilly and Company as a leading pharmaceutical manufacturer.[27] In 1922, the company teamed up with Charles Best and Sir Henry Dale of the National Institute for Medical Research in London in order to produce insulin commercially. By 1923, Eli Lilly and Company was able to mass produce insulin for worldwide distribution.[28]

The Era of Biologicals

Although the American pharmaceutical industry at the turn of the twentieth century was composed of mostly small manufacturing "houses" that specialized in manufacturing a few pharmaceutical products on a regional scale, a new set of challenges would change that. In the wake of the discovery of germ theory by Louis Pasteur (1822–1895) and Robert Koch (1843–1910) in the latter half of the nineteenth century, the fields of bacteriology and pharmacology were born. Pierre Paul Emile Roux (1853–1933), Pasteur's lab assistant, later discovered a method of mass producing diphtheria anti-toxin. Thus, pharmaceutical companies became interested in producing products that would treat infectious diseases. Starting in the 1890s a few pharmaceutical companies successfully began producing anti-toxins or "biological products" for diseases such as diphtheria and tetanus. Building on the research begun in France and Germany, the H.K. Mulford Company established a laboratory to produce anti-toxins in Philadelphia in 1894. By 1900, Mulford produced about a dozen anti-toxins including rabies, tetanus, and anti-streptoccocus.[29] Soon city boards of health began establishing laboratories to produce diphtheria anti-toxins as well. New York became the first in 1895 and its example was followed by Boston, Cincinnati, and St. Louis. After a series of children's deaths occurred in several cities due to faulty anti-toxin serum, the U.S. Congress passed the Biologics Control Act of 1902 which regulated and licensed serums and vaccines. The precursor to the U.S. Public Health Service was charged with establishing a standard anti-toxin for determining the strength of anti-diphtheria serum and for enforcement of the Act.[30] Parke, Davis, and Company and H.K. Mulford received the first licenses to produce biological products on a commercial basis.[31]

Until the outbreak of World War I, the U.S. pharmaceutical industry, much like the rest of the world, relied heavily on the scientific research performed in Germany. Most of the German pharmaceuticals sold in the US were manufactured in American plants under licensing arrangements and protected by U.S. patents.[32] The spoils of war changed all that. After the Treaty of Versailles, German assets were seized and auctioned by the Alien Property Custodian. One American company that benefited greatly from this was Sterling Products, Inc. founded in 1901 by William E. Weiss and A.H. Diebold. Sterling Products, Inc. was originally called Sterling Drug Incorporated and was founded in West Virginia.[33] Weiss and Diebold opened a factory and staked their fortunes on an analgesic they called "Neuralgine." The partners advertised aggressively and by diversifying their product line and acquiring other companies, they built their empire. By the end of World War I, Sterling was poised for the acquisition of Bayer Company, Incorporated of New York which they acquired in 1918 for $5,310,000. With this acquisition Sterling owned the rights to Bayer aspirin, the best selling pharmaceutical of all time. In 1970 alone, Sterling posted annual earnings of $594,412,000.[34] By 2007, Americans consumed 80 billion 300 milligram aspirin tablets annually, which retailed for about 1.5 cents a tablet.[35] In 1994, Bayer AG bought back the rights to Bayer aspirin.[36]

World War I had been a sobering experience for American pharmaceutical companies and prompted several of the larger ones such as Lilly, Merck,

Parke-Davis, Squibb, and others to open their own research institutes to conduct basic and translational research. Despite these efforts, most cutting edge pharmaceutical research still was conducted in Europe with the exception of insulin, which was discovered in Canada by University of Toronto researchers Frederick Banting and Charles Best. World War II once again caught the US unprepared and still reliant on foreign pharmaceuticals and chemical production.[37] A case in point was that by 1941 Japan controlled the world's supply of quinine to treat malaria. Winthrop company scientists scrambled to find out how to produce Atabrine, a synthetic alternative to quinine, using ingredients that were available in the US. Before American troops were deployed to regions where malaria was a threat they were issued Atabrine tablets.[38]

World War II and the Rise of "Big Pharma"

World War II proved to be a major watershed for the American pharmaceutical industry that catapulted it into an era of scientific research that would become the envy of the world. For example, on the eve of World War II the American pharmaceutical industry was still a collection of regional pharmaceutical "houses" that produced specialty products with a few national houses that produced a full line of products that still looked to Europe for research innovations. The immense challenges posed by the war called for an unprecedented amount of cooperation among government, corporate, and academic institutions. To cope with the medical challenges caused by the war, the federal government established the Committee on Medical Research of the Office of Scientific Research and Development (OSRD) to sponsor research that would have military applications. This partnership among government, corporate, and academic institutions laid the foundation for what became known as "big pharma." The OSRD sponsored research into antibiotics, antimalarials, blood products (dried plasma), and steroids.

The Emergence of Antibiotics

Perhaps the best known drug to emerge out of this era was penicillin. As noted in previous chapters, since prehistoric times mold and fungi have been used to treat infections. For example, the ancient Egyptians applied moldy bread to wounds. In modern times, Louis Pasteur and his colleague J. Joubert showed how an antibiotic could kill a bacterium in 1877.[39] Pasteur's and Joubert's discovery was followed by a number of other researchers on an experimental basis until, in 1928, the Scottish bacteriologist Alexander Fleming (1881–1955), of St. Mary's Hospital in London, noticed that *Penicillium* mold hindered the growth of staphylococci. Fleming published a paper on it and conducted a few preliminary experiments but abandoned the project. Howard Florey (1895–1968) an Australian researcher, Ernst Chain (1906–1979) a German biochemist, and a team of scientists at Oxford working with Merck isolated the antibiotic and tested it on animals and then humans successfully. With Britain under imminent attack, Florey brought his cultures to the United States and with the help of several pharmaceutical companies, including Merck, Pfizer, Squibb, Abbott, Lilly, Parke-Davis, Upjohn, Lederle, Reichel Laboratories, and Heyden Chemical produced mass quantities of the miracle drug. Commercial production by 1943 was 400 million units per month and by August 1945 production reached 650 million units per month.[40] In 1945, Fleming, Florey, and Chain shared the Nobel Prize for medicine for their astonishing contribution to humanity.

This singular pharmaceutical triumph ushered in the golden age of pharmaceuticals where modern medical miracles were expected, and in many cases delivered. Other antibiotics and sulfa drugs emerged after the war including Selman Waksman's streptomycin produced by Merck in 1945, chlortetracycline by Lederle in 1948, followed by chloramphenicol by Parke-Davis in 1949. Parke-Davis broke new ground in 1946 with Benedryl, an antihistamine. Since its inception in the late eighteenth century until the early 1960s, the American pharmaceutical industry concentrated on producing products that treated infectious diseases. After 1960, Big Pharma has focused its efforts toward producing drugs to combat chronic or hereditary illnesses. After the broad-spectrum antibiotics of the 1950s appeared, new classes of drugs including tranquilizers, antidepressants, antipsychotics, hormone replacement drugs, antacids, nsaids, proton pump inhibitors, anti-hypertensives, anti-cholesterol drugs, Cox II inhibitors, to name a few, came on the market. Also life-style drugs that had profound social implications emerged in the second half of the twentieth century, such as oral contraceptives and erectile dysfunction drugs.

Recent Trends

Since the 1980s, there have been four trends that have changed the course of pharmaceutical manufacturing: globalization, corporate mergers, biotechnology, and the search for the blockbuster drug (i.e., a patent drug that earns a company a half billion dollars annually). For example, the first of these blockbuster drugs was Smith, Kline, and French's Tagamet. Since the end of the Cold War in 1991 globalization has allowed pharmaceutical companies to venture into new emerging consumer markets in Asia, South America, and Africa. Globalization and trade agreements have accelerated the distribution of pharmaceuticals worldwide. Pharmaceutical companies sought acquisitions that would complement and expand their corporate expertise and grow their product lines and hopefully increase their market share of pharmaceutical sales on a global scale. In 1993, Merck entered new territory by acquiring Medco, a pharmacy prescription plan company. Lilly and SmithKlineBeecham followed suit, but after negative press all three companies divested these holdings by 1999. In recent years, Pfizer acquired the remnants of once prominent pharmaceutical firms including Wyeth, Upjohn, and Searle. Similarly, Merck acquired Schering. In the wake of the global recession of 2007, there is good reason to believe there may be even more corporate mergers and consolidation. It is hard to believe that as late as 1970 there were over 100 drug manufacturers in the United States.[41] Globalization ushered in an era of corporate mergers and acquisitions that rendered the pharmaceutical manufacturing landscape unrecognizable to all but the most astute financial and equity analysts. Gone were the once familiar pharmaceutical companies that for decades were closely associated with their product lines. Also gone were the days when pharmacists rose to become chief executive officers of major pharmaceutical corporations.[42]

Biotechnology

In search of new blockbuster drugs, research has shifted from organic chemistry increasingly toward molecular biology. Armed with basic research grants, university scientists began forming biotech companies in the mid-1970s, such as Genentech (1976), Biogen (1978), Amgen (1980), and Immunex (1981). Some companies, such as Genentech, became pharmaceutical companies in their own right by fusing their research engine with development and marketing departments to promote their hormonal products. Other biotech firms remained engaged in basic research and partnered with traditional pharmaceutical companies by licensing their intellectual property while relying on the drug companies to market their products.[43] Over the past few decades as several Big Pharma companies have shifted their focus from research and development to development and marketing, the research enterprise has devolved to smaller biotech firms and university settings. This has left a scenario where basic research and innovation has become diffuse and leaves Big Pharma companies as gatekeepers who possess the regulatory, development, and marketing expertise to select which drugs go to market and which ones do not. Moreover, since 1998 drug companies have engaged in direct to patient advertising and distribution, issuing samples and coupons for their drugs and bypassing wholesalers altogether.[44]

Chapter Summary

Since ancient Greek times, apothecaries have prepared drugs for patient consumption in ever increasing quantities that exploded with the emergence of the Industrial Revolution and pharmaceutical manufacturing in the early nineteenth century. As we have seen in the chapter, wars and scientific innovation have had a profound effect on how pharmaceuticals have been manufactured and distributed. From *Terra Sigillata* to Viagra, manufacturing whether done by hand or machine has been an integral part of the story of pharmacy.

Key Terms

Terra Sigillata

Glauber's salt

Ambrosius Gottfried
  Hanckwitz "Godfrey"

Antoine Baume

alkaloidal chemistry

John Farr

Abraham Kunzi

Heinrich Emanuel
  Merck

Stanislas Limousin

wafer cachet

Rosengarten and Sons

William R. Warner

John Wyeth

Edward Squibb

Frederick Stearns

Parke, Davis, and
  Company

Smith, Kline, and
  French

Eli Lilly

H.K. Mulford

Sterling Products

Chapter in Review

1   Describe the origins of the first trademarked drug *Terra Sigillata*.

2   Trace the evolution of various dosage forms prior to the Industrial Revolution.

3   Describe how national pharmacy societies got involved in manufacturing and distributing pharmaceuticals.

4   Describe Antoine Baume's contribution to pharmacy.

5   Assess the historical significance of alkaloidal chemistry on pharmaceutical manufacturing.

6   Describe Stanislas Limousin's contributions to pharmacy.

7   Assess the effects of warfare on the evolution of pharmaceutical manufacturing.

8   Trace the evolution of the American pharmaceutical industry from its start in Philadelphia to today.

9   Identify and assess the recent trends that have affected pharmaceutical manufacturing.

Notes

1 Andrea Tone and Elizabeth Siegel Watkins, eds., *Medicating Modern America: Prescription Drugs in History* (New York: New York University Press, 2007), 4.

2 Ibid.

3 George Bender, *Great Moments in Pharmacy* (Detroit, MI: Northwood Institute Press, 1966), 32–35.

4 George Urdang, "Retail Pharmacy as the Nucleus of the Pharmaceutical Industry," *Bulletin of the History of Medicine, Supplement*, No. 3 (1944), 325.

5 Ibid.

6 A.C. Wootton, *Chronicles of Pharmacy, Vol. I* (Boston, MA: Milford House, 1972), 260–262. David Cowen and William Helfand, *Pharmacy: An Illustrated History* (New York: Harry Abrams, 1990), 161.

7 Ibid. (Cowen and Helfand), 161.

8 George Urdang, "Retail Pharmacy as the Nucleus of the Pharmaceutical Industry,"
327.

9 C.J.S. Thompson, *The Mystery and Art of the Apothecary* (London: John Lane the Bodley
Head, Ltd., 1929), 327.

10 George Urdang, "Retail Pharmacy as the Nucleus of the Pharmaceutical Industry,"
330.

11 George Bender, *Great Moments in Pharmacy*, 103.

12 George Urdang, "Retail Pharmacy as the Nucleus of the Pharmaceutical Industry,"
331.

13 George Bender, Great Moments in pharmacy, 166–168.

14 David Cowen and Willam H. Helfand, *Pharmacy*, 167.

15 George Bender, *Great Moments in Pharmacy*, 166–168.

16 George Urdang, "Retail Pharmacy as the Nucleus of the Pharmaceutical Industry,"
337.

17 George Bender, *Great Moments in Pharmacy*, 94–95. Dennis B. Worthen, *Heroes of
Pharmacy* (Washington, DC: APhA Press, 2008), 60–61.

18 Glenn Sonnedecker, comp., *Kremers and Urdang's History of Pharmacy* (Madison, WI:
AIHP, 1976), 327.

19 John Swann, "The Evolution of the American Pharmaceutical Industry," *Pharmacy in
History*, Vol. 37 (1995), No. 2, 78.

20 George Urdang, "Retail Pharmacy as the Nucleus of the Pharmaceutical Industry,"
338.

21 John Swann, "The Evolution of the American Pharmaceutical Industry," 79.

22 John Swann, "The Evolution of the American Pharmaceutical Industry," 79.

23 Dennis B. Worthen, "The Pharmaceutical Industry, 1852–1902." In *American Pharmacy: A
Collection of Historical Essays*, edited by Gregory J. Higby (Madison, WI: AIHP, 2005), 58.

24 Glenn Sonnedecker, *Kremers and Urdang's History of Pharmacy*, 328. Dennis B. Worthen,
Heroes of Pharmacy, 203–204.

25 George Urdang, "Retail Pharmacy as the Nucleus of the Pharmaceutical Industry,"
338. Glenn Sonnedecker, *Kremers and Urdang's History of Pharmacy*, 328.

26 John Swann, "The Evolution of the American Pharmaceutical Industry," 80.

27 Gene R. McCormick, "Josiah Kirby Lilly, Sr.," *Pharmacy in History*, Vol. 12 (1970), No. 2,
57. Glenn Sonnedecker, *Kremers and Urdang's History of Pharmacy*, 331. George Urdang,
"Retail Pharmacy as the Nucleus of the Pharmaceutical Industry," 340.

28 George Bender, *Great Moments in Medicine* (Detroit, MI: Northwood Institute Press,
1965), 368.

29 Ramunas A. Kondratas, "The Biologics Control Act of 1902," in *The Early Years of
Federal Food and Drug Control* (Madison, WI: AIHP, 1982), 10. John Swann, "The
Evolution of the American Pharmaceutical Industry," 80.

30 George Griffenhagen, *150 Years of Caring* (Washington, DC: APhA, 2002), 132.

31 Dennis B. Worthen, "The Pharmaceutical Industry, 1902–1952," 62.

32 Dale Cooper, "The Trading With the Enemy Act of 1917 and Synthetic Drugs,"
Pharmacy in History, Vol. 47 (2005), No. 2, 48.

33 George Urdang, "Retail Pharmacy as the Nucleus of the Pharmaceutical Industry," 342.

34 Glenn Sonnedecker, *Kremers and Urdang's History of Pharmacy*, 332. Dennis B. Worthen,
"The Pharmaceutical Industry: 1902–1952," 64.

35 Diarmuid Jeffreys, *Aspirin: The Remarkable Story of a Wonder Drug* (New York: Bloomsbury, 2004), 270.

36 Ibid., 260.
37 Dale Cooper, "The Trading With the Enemy Act of 1917 and Synthetic Drugs," 48.
38 Dennis B. Worthen, "The Pharmaceutical Industry, 1902–1952," 64.
39 George Bender, *Great Moments in Medicine*, 372–385. George Bender, *Great Moments in Pharmacy*, 210–212.
40 Andrea Tone and Elizabeth Siegel Watkins, eds., *Medicating Modern America*, 2.
41 Stephen W. Schondelmeyer, "Recent Economic Trends in American Pharmacy," *Pharmacy in History*, Vol. 51 (2009), 121.
42 Dennis B. Worthen, "The Pharmaceutical industry, 1952–2002," 72.
43 Ibid., 71.
44 Stephen W. Schondelmeyer, "Recent Economic Trends in American Pharmacy," 121.

14 The Emergence of an American Institution

From the Corner Pharmacy to the Chain Drugstore, to the Big Box Department Store Pharmacy

What were the origins of the community pharmacy and how did it evolve?

Do you ever wonder how the modern pharmacy originated? One of the most universal of all human experiences is the trip to the local community pharmacy, a place where modern medicine and convenience store merchandising meet. When one thinks of the word "pharmacist," it conjures up an immediate and familiar image of a pharmacist behind a counter busily tending to the needs of his/her patients. Today community pharmacies take many shapes and forms from the corner drugstore, the pharmacy in the supermarket, the chain pharmacy, or the big box department store pharmacy. This chapter is about the origin and evolution of the great American institution, the community pharmacy.

What was the first drugstore in America? This question has created quite a debate among historians of pharmacy due to the imprecise nature of the question and the lack of definitive historical evidence. Of course, Native Americans had an elaborate system of medicine that included a wide array of herbs that, over time, were found to be medicinally useful and adopted by Europeans and Americans and included in their respective pharmacopoeias. Their system of medicine did not "sell" drugs in the European sense and therefore did not operate commercial drugstores. As discussed in Chapter 9 of this book there were several contenders who have a claim to have started the first drugstore in North America. The historian George Bender recounts the case of Louis Hebert (1575?–1627?) who abandoned the comfortable, but predictable life of a Parisian apothecary to settle in Nova Scotia and eventually Quebec to practice pharmacy among the settlers and Native Americans in the seventeenth century.[1] The venerable Kremers and Urdang textbook noted that there was an herbal garden, mostly growing imported herbs, in New Orleans as early as 1724. Sieur Dameron who was in charge of the herbal garden also operated a laboratory that produced medicines for the local garrison and hospital. Ursuline nuns ran the hospital pharmacy until the Spanish took control in 1763.[2] Similarly, a French colonist in Louisiana (which came under the control of Spain for 28 years) named Jean Peyroux passed his examination in 1769 and was certified under Spanish law to open a pharmacy.[3] The historical record became murky during this era as Louisiana passed from French to Spanish and back to French control before it was purchased by the United States in 1803.

French and Spanish law continued to have considerable influence over regulating the practice of pharmacy. At least two candidates, François Grandchamps and Louis J. Dufilho, Jr., became the first apothecaries to earn licenses to practice in the United States in 1816.[4] Dufilho opened an apothecary shop in New Orleans in 1823 and practiced for many years. His store has been converted into a pharmacy museum in the French Quarter of New Orleans and makes the claim that Dufilho was America's first licensed pharmacist. Sadly, although Louisiana had 130 registered pharmacists by 1847, in 1852 the state succumbed to the prevailing spirit of laissez-faire and repealed all of its laws that regulated pharmacy, medicine, and dentistry. Louisiana and several other Southern states that had early pharmacy regulations only to repeal them in the 1850s did not reinstate them until the 1870s and 1880s.[6]

Ironically, graduating from a pharmacy school and passing a board exam would not be required by any state until New York passed such a law in 1904.[5]

Pharmacy in New Amsterdam

Shortly after the Dutch settled in New Amsterdam in 1626 there were at least two surgeons Herman Meynders van den Boogaert and Gysbert van Imbroch (d. 1665) who began to practice. Boorgaert was in charge of the stores and provisions for New Amsterdam (later New York City) and later Fort Orange (later Albany, New York). Van Imbroch opened a shop in New Amsterdam in 1653, perhaps a drugstore, perhaps not. He did operate a general store in Wildwyck from 1663–1665, which might have been the first drugstore in North America. This general store in what is today Kingston, New York may have been the prototype of what became the "doctor's shop" in which patients visited with the physician/surgeon and then were sold prescriptions in the same location. If the practice could support them, sometimes physicians employed drug clerks to prepare and dispense drugs to patients. Over time, some of these drug clerks opened their own shops, usually in cities that could support such a specialized operation. The doctor's shop became the predominant and earliest prototype for the drugstore as we might know it.[7]

Pharmacy in the British North America

Not to be outdone by the French or Dutch, the British sent Bartholomew Gosnold to the New World and he returned to England from what later became Massachusetts with a load of sassafras.[8] This was about the same time that Sir Walter Raleigh returned from the coast of what later became North Carolina with various herbs including sassafras, china root, and other products. A year after Jamestown colony was founded in Virginia, two apothecaries named Thomas Field and John Harford arrived.[9] Later, Thomas Wharton established and operated an apothecary shop in Williamsburg, Virginia in the mid-1730s. One of the claimants to have operated the first pharmacy in America was the physician/apothecary Hugh Mercer (1725–1777) who opened a shop in Fredericksburg, Virginia in 1764. George and Martha Washington were regular patrons of Mercer's shop.[10] (See Chapter 9.)

After the Pilgrims landed in Plymouth and the Puritans settled in Massachusetts Bay Colony in 1620 and 1628, respectively, the first pharmacy practiced was performed by none other than the Bay Colony's first governor, John Winthrop. Prompted by necessity, Winthrop learned about native herbal remedies from the Native Americans and ran a primitive apothecary shop from his home, preparing herbal remedies for his fellow colonists. The first record of a trained apothecary to arrive in Massachusetts Bay Colony was Robert Cooke who came to Boston in 1638. The earliest owner of an apothecary shop who had some formal training was a physician named William Davis (Davice) (*c.*1617–1676) who established a shop in Boston in 1646. Once the British gained control of most of North America during the 1760s, for better or worse, American pharmacy would be more heavily influenced by the rather chaotic British system rather than the more clearly defined Continental European system.

A great deal of colonial pharmacy in North America was driven by the colonial mercantilist system in which the mother country monopolized trade and imported raw materials from the colonies, selling back finished products to the colonists. Imported drugs were sold not only in apothecary shops but as sideline products in general stores, tailor shops, coppersmith shops, and post offices. American Revolution heroes and villains alike, such as Paul Revere and Benedict Arnold, sold these imported drugs in shops they owned. Perhaps the best known and earliest of what became the model

Table 14.1 Early Pharmacies

Owner	Year	Location	Name of Shop
Christopher Marshall	1729	Philadelphia, PA	
Thomas Wharton	1735	Williamsburg, VA	
Sylvester Gardiner	1775	Boston, MA	"Unicorn and Mortar"
G. Duykinck	1769	New York, NY	"Looking Glass & Druggist Pot"
Richard Speaight	1776	New York, NY	"Elaboratory"
Evan Jones	?	Philadelphia, PA	"Paracelsus' Head"
Phillip Moser	?	Charleston, SC	"Man and Mortar"
William Brisbane	?	Charleston, SC	"Eagle"[11]

for eighteenth-century American pharmacy that bridged the Colonial and Early Republic eras was Christopher Marshall's wholesale pharmacy, established in Philadelphia in 1729, which remained in the family's hands until it was sold in 1825. Marshall's store began as a wholesale operation preparing medicine chests for physicians, surgeons, apothecaries, and large landowners employing drug clerks in the back end laboratory in a warehouse to prepare the medicines. Over time, as a sideline, Marshall opened the front end of his operation to walk-in customers to sell retail prescriptions to them. The front end of these retail shops were used to take advantage of the natural light and to allow patients to watch as their prescriptions were being compounded, lending an artisanal quality to the apothecary's persona.[12] Marshall's "drugstore" became a training ground for many apothecaries and laid the foundation for the Philadelphia College of Pharmacy in 1821 with his son Charles serving as its first president. These wholesalers in the larger cities who also opened retail operations came to be known as "druggists." They ran these warehouses or storehouses of drugs, which became known as "drugstores."

In the wake of the War of 1812, medical education evolved from apprenticeship to formal medical schools where physicians learned to write prescriptions rather than to compound them. To be sure, doctor's shops remained, but specialized apothecary shops began to appear, one of the best known being that of Elias Durand (1794–1873) who was perhaps the first apothecary to have a soda fountain in his shop. He also developed a bottling apparatus for soda. Durand, a French immigrant, acted on his chemical training by conducting experiments in his shop and whetted the scientific curiosity of many aspiring apothecaries, the most famous being William Procter, Jr. In fact, immigrants from France had a key influence on advancing pharmacy practice in New Orleans. Similarly, German immigrants exercised a profound influence in shaping the way pharmacy was practiced in New York, St. Louis, and parts of eastern Pennsylvania, often elevating practice standards by virtue of the specialized training they brought with them from Europe. By the 1830s, these pharmaceutical scientists became known as "pharmaceutists" by virtue of their formal education and specialized knowledge of the revolution in alkaloidal chemistry that produced drugs such as morphine, caffeine, strychnine, and other substances.

While the lines of medical and pharmacy practice had been blurred from the start in the United States, this state of confusion had not caused any real tensions between physicians and apothecaries because there were so few of them. Geographically, the new nation was large and the market for medical services was growing, but by the 1840s American medical schools were flooding the market with new graduates. This created unprecedented competition and many physicians began selling drugs as a sideline in addition to providing their traditional services. Apothecaries retaliated and began "counter prescribing," a practice where they diagnosed patients on the spot and sold them remedies, including patent medicines, which aggravated an already tense situation. This tense state of affairs served as a catalyst in the formation of the American Medical Association in 1847 and consequently the American Pharmaceutical Association in 1852. A first order of business for both organizations was to call for clear role definitions about what their practitioners precisely were expected to do. Unfortunately, this tension between apothecaries and physicians remained until legislation settled it.

Antebellum Drugstores

Due to population growth, the American market could support specialization and apothecary shops emerged. In the East Coast cities, in addition to the front end prescription business these shops sold patent medicines, toiletries, dyes, and in many cases books. On the frontier, the drugstores were configured more as general stores with a prescription department and a wholesale department. As general stores, these shops sold glass, paint, dyes, putty, snuff, potash, liquor, and the necessities of life. One obstacle that hindered the emergence of the specialized apothecary shop on the frontier was the fact that the stores were compelled to operate on a barter system with their customers, often accepting furs and other raw materials as payment in what was known as "store pay." As frontier areas were settled, more specialized apothecary shops emerged.[13]

The Golden Age of the American Drugstore

Most historians agree that the era after the Civil War witnessed the birth of the American drugstore. The birth of large-scale pharmaceutical manufacturing that had accelerated during the war caused a dramatic transformation in how apothecaries practiced and how they operated their shops. With the war serving as a catalyst, the full effects of the Industrial Revolution transformed how goods were produced and distributed. Products could be mass produced in remote factories, taking advantage of cheap labor, and then shipped by rail to wholesalers and shops for sale. This had profound implications for pharmaceutists and their shops. Manufactured drug ingredients replaced the need for pharmaceutists to produce these in the front end of their shops. With the front end rendered superfluous, pharmaceutists moved their prescription laboratories to the back end of the store and often elevated and screened off the area where they prepared prescriptions, lending an air of mystery to their work.

In contrast to their European counterparts, American pharmacists relied on selling sideline products in their stores to remain profitable. According to Gregory Higby, between the Civil War and World War II the average pharmacist filled an average of 6 to 12 prescriptions per day.[14] For example, in the 1930s the average drugstore received only eight prescriptions per day but this average varied by region.[15] The American prescription drug business would not become profitable for most drugstores until after World War II, especially with the introduction of antibiotics to the public.[16] With the pre-scription department in the back of the store, pharmacists seized the oppor-tunity to sell other goods and services in the front end of their store, making them something akin to small emporiums. Drugstores began selling various sundries including candy, newspapers, magazines, and toiletries.

The Introduction of the Drugstore Soda Fountain

One innovation that transformed the apothecary's shop into the drugstore was the introduction of the soda fountain. Since Joseph Priestly's and Tobern Bergman's experiments in the 1770s which produced carbonated water, soda fountains had come a long way and provided customers with a new reason to visit their local drugstore.[17] Jacob Baur, a pharmacist, experimented and perfected carbon dioxide tanks and in 1888 established the Liquid Carbonic Company. By 1900, Baur began producing and selling his soda fountains complete with an instruction and recipe manual, making it simple for any-one to set up their own business. Mass-produced syrups including Coca-Cola, Ward's Orange Crush, Cherry Smash, Orange Julep, Hires' Root Beer, Dr. Pepper, and others became customer favorites associated with the American drugstore. As discussed in Chapter 10, pharmacists had a special knack for developing unique flavors for these syrups.

The Invention of the Milkshake

Other soda fountain treats that came into vogue after the Civil War were milk-based drinks. The first of these was Koumyss, a drink made from sour cow's milk, fresh milk, and sugar. This drink had its origins in the ancient Asian drinks called kumiss made from mare's milk and from kefir made from camel's milk. Another drink popular during the late Victorian era was malted milk. William Horlick from Racine, Wisconsin, claims to have coined the term "malted milk," which was made from wheat and barley extracts and dried milk.

Did you know that pharmacists had something to do with the origin of the "cocktail"? During the late eight-eenth century New Orleans apoth-ecary, Antoine Peychard, dispensed tonics made from cognac and his own Peychard's bitters. This product was served in an eggcup, which in French was known as a *coquetier*, giving us the modern term "cocktail."[19]

He sold this in powder form as a dietary supplement and it was not long before some curi-ous druggists began mixing it with ice cream at their soda fountains. Milk-based drinks can be traced back to the 1880s, but they really became popular during the 1920s.[18] One of the great sensations in soda fountain history was created by a Walgreens Soda

Fountain Manager, Ivar "Pop" Coulson, who created the double-rich chocolate malted milk in the Chicago loop store in 1922. This milkshake created a minor sensation in Chicago for Walgreen and helped to shape the young company's identity for innovation and value to the customer.[20]

Made during the patent medicine era, many of the early soda fountain syrups contained cocaine and caffeine, which caused many a customer to saddle up to their local soda jerk literally asking for "a fix." This practice continued until the federal Harrison Narcotic Act was passed in 1914, which banned the use of opiates or cocaine in over-the-counter products and placed them under strict controls.[21] In part, this law was prompted by infants becoming addicted to opiates due to their use as a popular treatment for teething pain. The soda fountain weathered this storm by serving tasty soft drinks and ice cream treats. With the rise of the temperance movement and its success in getting the Volstead Act passed, the Prohibition Era from 1920 until 1933 saw the number of soda fountains increase dramatically as bars closed. The 1920s through the 1950s was a golden age for the soda fountain and the American drugstore, with many becoming small neighborhood restaurants. The Prohibition Era also enhanced the prescription business for drugstores as many consumers received prescriptions for "medicinal" alcohol from their physicians.[22]

Price Cutting

American pharmacy developed in a largely unregulated laissez-faire atmosphere of what the American market would bear in the nineteenth century. This created "occupational fluidity" in which any enterprising individual could open and operate a drugstore with a minimal degree of training well into the 1870s, prior to the era of state board regulations. From the start, the drug business was very competitive and one of the results of laissez-faire policies was the phenomenon of price cutting. One of the people who pioneered this practice was an English immigrant named Thomas W. Dyott (*c.*1775–1850). Dyott opened his Drug Warehouse in Philadelphia about 1807 where he manufactured and sold a full line of remedies. From the start, he posted signs promising the public "cheap drugs," undersold his competition, and advertised relentlessly. His success was only tempered by the national financial crisis of the 1830s.[23]

Nothing succeeds like success and Dyott's example was followed by the George A. Kelly Company, which engaged in price cutting in Pittsburgh around 1860. The company owned four drugstores that offered the public cut-rate drugs. The price-cutting situation was exacerbated by the mass production of drugs by pharmaceutical manufacturers after the Civil War. Manufacturers and wholesalers offered volume discounts for bulk purchases which encouraged further price cutting, especially by large department stores that were being established in the 1880s. In the 1880s the price cutters included Dow in Cincinnati, Evans in Philadelphia, Jacobs in Atlanta, and Robinson in Memphis. George Ramsay established Hegeman Company which brought price cutting to New York.[24] There were a few attempts made to curtail the excesses of price cutting by NARD, NWDA, and the Proprietary Association of America, but these efforts met with little success. Some progress was achieved during the Great Depression Era of the 1930s, but price cutting continues to be a challenge for community pharmacies amid a free enterprise system. Beginning in the 1960s, third party payers such as health insurers and government programs such as Medicare and Medicaid and the creation of pharmacy benefit managers have added pressure for pharmacies to engage in further price cutting.

Chain Drugstores

The concept of the chain drugstore began in Britain with chains such as Boots, Ltd. After the Civil War a few chains emerged including Cora Dow of Cincinnati, Hall and Lyon in Providence, Hegeman and Company in New York, and Charles B. Jayne in Boston. These chains were followed by Economical-Cunningham Drugstores, Read Drug and Chemical Company (now Rite Aid), Hook Drug, Peoples Drug, and Thrifty Drugstores.[25] Nonetheless, the modern chain drugstore concept developed in earnest by the vision and work of two enterprising men: Charles R. Walgreen and Louis K. Liggett.

Charles Rudolph Walgreen, Sr.

Charles R. Walgreen, Sr. (1871–1939) bought his first drugstore in Chicago in 1901 from his employer with borrowed money and parlayed that purchase into one of the greatest success stories in American business. Born in Rio, Illinois in 1871, Walgreen arrived in Chicago in 1893 and bounced around the city working as a drug clerk. Self-taught, Walgreen passed the Illinois State Board Exam and became a pharmacist in 1897. He enlisted in the Illinois National Guard and served in a hospital during the Spanish-American War in Cuba. He contracted malaria, nearly died, but recovered and returned to Chicago where he bought Isaac Blood's drugstore for $6,000 and found his life's calling. Walgreen introduced the hallmarks of his stores; clean, shiny floors and clear glass display cases for the merchandise. He learned about merchandising by experimenting with a special display table in the middle of his store to sell a load of pans that he bought from a salesmen for a good price. Although selling pans had little to do with pharmacy, Walgreen sold the pans within a few days, earning a handsome profit and also attracting new customers to his store. Walgreen made that the linchpin of his business plan; sell a wide variety of products that other drugstores did not sell to give customers more reason to visit his stores.

Even though his soda fountain was popular during the summer, he closed it during the winter months. With the help of his wife Myrtle, Walgreen established a food service that offered a limited menu of sandwiches, soup, and pie that became a successful part of his business, once again giving customers another reason to visit his store. Walgreen also devoted a large part of his budget to newspaper advertising. Walgreen's spirit of entrepreneurial innovation lived on in his company. Over time, his company introduced self-service shopping, lower cost store brand merchandise, superstores, one hour photo developing, drive through pharmacy, computerized prescription services, and recently, urgent care clinics. In fact, next to the United States government, Walgreens is the world's second largest satellite user, using a database that connects its 8,217 stores across 44 states.[26] In 1937, Walgreens opened its first "superstore" in Miami. During the 1950s the company experimented with customer self-service in its stores in the Midwest and by 1953 became the nation's largest self-service chain. While many of its competitors, including Montgomery Ward, A&P, Rexall, and others, have vanished, Walgreens continues to thrive. In June 2012, Walgreen Company purchased a $6.7 billion stake in the European drugstore and beauty products retailer chain Alliance Boots, giving it global clout. By its acquisition of Alliance Boots' 2,400 stores, mostly located in the United Kingdom, the Walgreen Company has completed the historical circle of drugstore chain history. Alliance Boots was one of the first drugstore chains in the world, founded by Sir Jesse Boot in the nineteenth century.[27]

Louis K. Liggett and Rexall

Another key figure in the saga of the chain drugstore was the entrepreneur Louis Kroh Liggett (1875–1946), who parlayed his fortune selling the patent medicine "Vinol" into one of the largest manufacturing, wholesale, and marketing companies in American history. Unlike Charles Walgreen, Liggett was a businessman and not a pharmacist. Still, Liggett was an innovator whose time had come at the dawn of the twentieth century. In the atmosphere of

Table 14.2 Walgreen Company Growth[28]

Year	Number of Stores
1984	1,000
1994	2,000
2000	3,000
2004	4,000
2005	5,000
2007	6,000
2011	7,800
2014	8,217

price cutting that dominated the drug business, Liggett believed the key to success was in offering factory prices directly to one pharmacist in a town and controlling the distribution of these company-made products. In 1902, Liggett persuaded 40 investors from around the United States to invest and founded the United Drug Company with him serving as general manager and secretary. The new company was about to market its products under the name "Saxona," but W.T. Wilson, one of Liggett's office assistants suggested the name "Rexall" or the "king" of all products. Another explanation was that Rexall was an acronym meaning prescriptions or "RX for all."[29] So, the Rexall brand was born. The company got off to a profitable start and in 1904 Liggett and his company opened their own drugstore that contradicted the initial franchise-only model. By 1905, Liggett had entered the candy and tobacco business by founding the National Cigar Stands Company that gave discounts on its products to United Drug Company franchisees. In 1916, Liggett established the L.K. Liggett Company that owned 152 stores and had agreements with 6,000 other franchise stores.[30]

Liggett's dream of a drugstore empire knew no bounds. On a visit to England in 1912, Liggett organized 126 investors to establish a company along the lines of the United Drug Company he had founded in the United States. In 1920, Liggett bought Boot's Drug Company, which had been established by the venerable Sir Jesse Boot himself. That same year, United Drug Company opened its own manufacturing plant in St. Louis. On the eve of the Great Depression Liggett had amassed nearly 1,000 stores in Britain and the United States, but had done so during uncertain economic times. As early as 1921, United Drug Company's stock price had fallen by $60 per share and despite Liggett's call for the company to pull together and stand firm, the company faltered. While United Drug remained on solid financial ground, L.K. Liggett Company failed in 1933, prompting a reorganization and absorption of both companies under a new entity called Drug Incorporated. Desperate, Liggett negotiated a difficult deal to sell his British assets and earned a profit of $22 million that allowed him to buy back the Liggett drugstore chain in the United States.

Despite the Great Depression, innovative entrepreneur that he was, Liggett established a pharmaceutical research division at United Drug in 1935. A year later, Liggett wanted to show the vitality of his company and in a flamboyant, but effective promotional move he introduced the Rexall Train. The air-conditioned train consisted of 12 cars including a modern pharmacy complete with a soda fountain, a model of Rexall's research facilities, a model of Rexall's manufacturing plants, and living quarters for Liggett. The streamlined blue train logged over 30,000 miles, stopping for one-day business meetings and exhibitions in 150 cities. Upon his retirement in 1941, the United Drug Company had weathered the Great Depression and owned 600 drugstores with about 8,000 franchisees selling 5,000 products with 16,000 employees.[31] The Rexall name lived on until the age of arbitrage and acquisitions of the

1980s changed its emphasis from manufacturing to distributing products. In 1985, Rexall's assets were sold to a Florida-based investment group known as the Rexall Group. W.T. Wilson's grand moniker "Rexall" was sold in 1989 to an investment group known as the Sundown Company.[32] Today, Rexall brand products are still being sold in select outlets throughout the United States, but on a far smaller scale than during its glory days several decades ago.

The Rise of Supermarket and Big Box Retail Pharmacies

Since the 1960s several supermarket chains have operated pharmacy departments in their stores, which has posed significant challenges for the traditional drugstore. In 2008, these supermarket pharmacies constituted 16.6 percent of all pharmacies in the United States. Similarly, since the 1970s several discount department stores have opened pharmacy departments in their stores. National "Big Box" or mass merchandise pharmacies such as K-Mart, Target, and Walmart account for 13 percent of all pharmacies in the United States.[33] More recent discount warehouse chains such as Sam's Club also feature in-store pharmacy departments. Since the 1980s with the rise of pharmacy benefit management companies, mail-order pharmacy for maintenance drugs has made its presence felt, offering discounts for 90-day supplies of drugs. This phenomenon was followed by the internet pharmacies established in the 1990s. Convenience and price cutting continue to be the leading motifs in the world of community pharmacy.

Chapter Summary

The American community pharmacy has come a long way from the small doctor's shops of colonial times to the ubiquitous and convenient chain store pharmacies of today. Long gone are the days of the pharmaceutist who personally compounded every drug he sold. Pharmaceutical manufacturing, especially after the Civil War increasingly rendered compounding obsolete, compelling the pharmacist to find other ways to stay in business. By the 1860s, drugstores became general stores and emporiums that relied on selling sideline products of every sort in order to stay in business. Some drugstores embarked on installing soda fountains and later featured restaurants. From the 1890s through the early 1970s the drugstore soda fountain served as a community gathering place that inspired a folklore of its own. By the early years of the twentieth century, entrepreneurs led by Charles Walgreen and Louis Liggett established the chain pharmacy that has become the dominant form of American community pharmacy. While chain pharmacies still dominate, American community pharmacy competitors and market forces have given rise to significant challengers. Supermarket pharmacies, big box retail department store pharmacies, internet pharmacy, and mail-order pharmacy are transforming the way Americans shop for their drugs in the twenty-first century.

Key Terms

Doctor's shops	Christopher Marshall	Louis K. Liggett
François Grandchamps	Elias Durand	Rexall
Louis J. Dufilho, Jr.	William Procter, Jr.	United Drug Company
Gysbert van Imbroch	American	Big Box Retailers
Hugh Mercer	Pharmaceutical	Walmart
John Winthrop	Association	Target
William Davis	Thomas W. Dyott	K-Mart
Paul Revere	price cutting	
Benedict Arnold	Charles Walgreen, Sr.	

Chapter in Review

1   Describe the colonial origins of the American community pharmacy.
2   Describe what a "doctor's shop" was.
3   Describe the significance of Christopher Marshall's apothecary shop in Philadelphia on the development of American community pharmacy.
4   Trace the origin of the tension between physicians and apothecaries and describe how these tensions over conflicting roles were resolved.
5   Trace the role soda fountains played in American community pharmacy. Discuss how this changed over time.
6   Describe how "price cutting" has affected American community pharmacy.
7   Trace the origin of the American chain pharmacy and its prospects for the twenty-first century.

Notes

1 George Bender, *Great Moments in Pharmacy* (Detroit, MI: Northwood Institute Press, 1966), 72.
2 Glenn Sonnedecker, comp., *Kremers and Urdang's History of Pharmacy* (Madison, WI: AIHP, 1976), 151.
3 Ibid., 149.
4 Ibid., 158. Laurel Dorrance, "An Examination of the Education and Licensing of Pharmacists in Early Louisiana, 1718 to 1816: The Significance of Louis Joseph Dulfilho, Jr.," *Pharmacy in History,* Vol. 53 (2011), Nos. 2 & 3, 80.

5 Robert Buerki, "Pharmaceutical Education, 1902–1952." In *American Pharmacy: A Collection of Historical Essays*, edited by Gregory Higby (Madison, WI: AIHP, 2005), 44.

6 Glenn Sonnedecker, David L. Cowen, and Gregory Higby, eds., *Drugstore Memories: American Pharmacists Recall Life Behind the Counter, 1824–1933* (Madison, WI: AIHP, 2002), 2–3.

7 Glenn Sonnedecker, *Kremers and Urdang's History of Pharmacy*, 151.

8 George E. Osborne, "Pharmacy in British Colonial America." In *American Pharmacy in the Colonial and Revolutionary Periods*, edited by George Bender and John Parascandola (Madison, WI: AIHP, 1977), 7.

9 Ibid., 6.

10 William H. Helfand, *Potions, Pills, and Purges: The Art of Pharmacy* (Madison, WI: AIHP, 1995), 9.

11 Glenn Sonnedecker, *Kremers and Urdang's History of Pharmacy*, 158.

12 Glenn Sonnedecker, David L. Cowen, and Gregory Higby, *Drugstore Memories*, 4.

13 Glenn Sonnedecker, *Kremers and Urdang's History of Pharmacy*, 293.

14 Glenn Sonnedecker, David L. Cowen, and Gregory Higby, *Drugstore Memories*, 69.

15 Ibid., 7.

16 Ibid., 68.

17 Ann Cooper Funderburg, *Sundae Best: A History of Soda Fountains* (Bowling Green, OH: Bowling Green State University Popular Press, 2002), 6–7.

18 Ibid., 50–51.

19 David L. Cowen and William H. Helfand, *Pharmacy: An Illustrated History* (New York: Harry Abrams, 1990), 188.

20 John U. Bacon, *America's Corner Store: Walgreens' Prescription for Success* (Hoboken, NJ: John Wiley & Sons, Inc., 2004), 230. Herman Kogan and Rick Kogan, *Pharmacist to the Nation: A History of Walgreen Co.* (Deerfield, IL: Walgreen Company, 1989), 68–69.

21 Joseph F. Spillane, "The Road to the Harrison Narcotics Act: Drugs and Their Control, 1875–1918." In *Federal Drug Control: The Evolution of Policy and Practice*, edited by Jonathon Erlen and Jospeh F. Spillane (New York: Pharmaceutical Products Press, 2004), 1–21. W. Stephen Pray, *Nonprescription Product Regulation* (Binghamton, NY: Pharmaceutical Products Press, 2003), 78.

22 Ibid. (Pray), 56.

23 Glenn Sonnedecker, *Kremers and Urdang's History of Pharmacy*, 295.

24 Ibid.

25 Ann Cooper Funderburg, *Sundae Best*, 127.

26 *St. Louis Post-Dispatch*. Business Section, 22 January 2012, E1.

27 *St. Louis Post-Dispatch*, Business Section, 20 June 2012, A10.

28 John Bacon, *America's Corner Store*, 231. Walgreens.com. *St. Louis Post-Dispatch*, 22 January 2012, E1.

29 Ann Cooper Funderburg. *Sundae Best*, 128.

30 Mickey C. Smith, *The Rexall Story: A History of Genius and Neglect* (New York: Pharmaceutical Products Press, 2004), 8.

31 Ibid., 12.

32 Ibid., 153.

33 Stephen W. Schondelmeyer, "Recent Economic Trends in American Pharmacy." In *Pharmacy in History*, edited by Gregory Higby (Madison, WI: AIHP) Vol. 51 (2009), No. 3, 119.

15 The Impact of Federal Legislation on the Practice of Pharmacy

What factors caused the American drug market to evolve from one of the least regulated markets in the world to become one of the most regulated and safest drug markets in the world?

Few things have affected the practice of pharmacy in the United States as profoundly as federal legislation, or more to the point, the lack of it in the nineteenth century. The very first article of the U.S. Constitution in 1787 granted Congress the power to grant patents and copyrights to inventors and authors to profit from their work.[1] Congress passed the first patent Act in 1790 and the first modern patent law was passed later, in 1870. These patent laws gave rise to a proliferation of medicines of various standards of quality known as "patent medicines." More precisely, patent medicines were really proprietary medicines, because in order to obtain a patent, manufacturers would have been obliged to reveal the ingredients of their products, which they were not interested in doing. For the most part, all patent medicines were made from the same herbs, fruit juices, alcohol, and sometimes contained narcotics such as opiates or cocaine. Instead, these medicine makers patented their labels, bottles, and trademarks in order to claim their market niche. From the beginning of the Republic until 1906, patent medicine makers were free to make sensational therapeutic claims for their untested and unproven products.[2]

The Drug Importation Act of 1848

The American Medical Association (AMA) and the American Pharmaceutical Association (APhA) were founded in 1847 and 1852, respectively, in Philadelphia, in part to educate the public about the dangers inherent in patent medicines. Both organizations embarked on publicity campaigns and called for legislation to regulate the production and sale of patent medicines, but faced stiff opposition from manufacturers. The first such federal legislation was the Drug Importation Act of 1848, which prohibited "the importation of adulterated and spurious drugs and medicines."[3] Amid a prevailing belief in laissez-faire economic theory Congress was reluctant to act until reports surfaced about soldiers serving in the Mexican War who were harmed by adulterated Peruvian bark, imported from overseas, used to treat malaria.[4]

The Drug Importation Act, signed into law by President James K. Polk, called for the training and appointment of drug examiners at the main ports of the United States to inspect incoming drugs for quality, purity, and fitness for medical use. While the law seemed to work in theory, in practice many of the port examiners were political appointees whose political loyalties in many cases trumped their degree of expertise. Also foreign drug manufacturers learned to circumvent the law by simply shipping their products to another port if their goods failed the inspection at the first port. One egregious case of an attempt to import a shipment of foreign drugs led the College of Pharmacy of the City of New York in 1851 to call for a national meeting of pharmaceutists to create a set of standards for the imported drugs according to the standard pharmaco-poeia and dispensatories. This meeting in 1851 ultimately led to the creation of the APhA in 1852 in Philadelphia. From its start the APhA dedicated its efforts toward drug safety, be it imported drugs or patent medicines.[5]

Biologics Control Act of 1902

In the wake of germ theory, vaccines and serums (biologi-cal products) gained promi-nence as an important public health concern in prevent-ing infectious diseases, namely diphtheria and tetanus. In the United States the H.K. Mul-ford Company founded the first laboratory to produce anti-toxins in 1894 in Phila-delphia. Led by the New York Board of Health Laboratory in 1895, other city boards of health began producing diph-theria anti-toxins and serums. Unfortunately, in 1901 at least five children died from tetanus in St. Louis after receiving contaminated diphtheria anti-toxin which had been manufactured by the city's health department. Other tragedies around the nation prompted Congress to pass the Biologics Control Act of 1902 that regulated and licensed serums and vaccines. The precursor to the U.S. Public Health Service was charged with regulatory responsibility that included the licensure and inspection of facilities that produced biological products.[7]

I Can't Believe It's Not Butter? The Butter Wars

While the sale of patent medicines flourished, the sale of oleomagarine came under Congressional scrutiny. Oleomagarine was invented in 1869 as a substitute for butter that was cheaper and also had a longer shelf life than but-ter. With aggressive advertising, oleo-magarine gained steady market share and thus butter makers appealed to Congress for relief in 1886. Congress obliged and passed a law placing a tax and other regulations on oleomagarine that remained until 1950.[6]

The Pure Food and Drug Act of 1906

For many years patent medicines and the dangers they posed to the public became subjects of debate at the annual meetings of the APhA and AMA. Both organizations passed resolutions calling for federal legislation to regulate these nostrums. The APhA and AMA were encouraged by Britain's 1875 law that regulated foods and drugs. Between 1879 and 1906 about 100 bills had been proposed and, due to powerful opposition from patent medicine makers, had failed until the Pure Food and Drug Act was passed. Nonetheless, there was hope. By 1895 about half of the states had passed food and drug safety laws, but the laws were inconsistent and could be circumvented by manufacturers selling a product in another state.[8]

A key figure in the fight to promote drug and food safety was Dr. Harvey Washington Wiley (1844–1930). Wiley was born on October 18, 1844, in a log cabin in Indiana which served as a stop on the Underground Railroad for slaves seeking their freedom. Wiley began his studies at Hanover College, but answered the call and served in the Union army during the Civil War. He earned his medical degree from Indiana College in 1871 and later a bachelor of science from Harvard in 1874. He joined the faculty as a chemist when Purdue University was founded in 1874, where he was censured for riding a high-wheeled bicycle and playing baseball with students. After studying in Berlin for a year with Eugene Sell, he returned to the United States where he analyzed sugars and syrups for the Indiana State Board of Health. Wiley had found his calling in promoting food and drug safety.[9]

In 1882, Wiley became the chief chemist of the U.S. Bureau of Chemistry in the Department of Agriculture in Washington, DC, where he worked on trying to find a safe additive to preserve various foods. Wiley proved to not only be a talented chemist, but also a gifted coalition builder by marshalling the forces of the AMA, APhA, civic groups, investigative "muckraker" journalists, and state food and drug safety officials. For example, a journalist for *Collier's Weekly*, Samuel Hopkins Adams, wrote an investigative series of reports from 1905–1907 entitled, "The Great American Fraud." The most sensational story was that of Dennis Dupuis, aka "Dr. Rupert Wells, MD," of the Postgraduate College of Electrotherapeutics of St. Louis. Dupuis claimed his tonic called "Radol" cured cancer and offered a treatment regimen for the handsome sum of $15 per month. Of course, Dupuis' medical degree, college, and tonic's claims were pure fabrications.[10]

Perhaps the best known and most powerful call for reform came in the form of Upton Sinclair's (1878–1968) novel, *The Jungle*, which exposed the unsanitary practices of the meat-packing industry and prompted a public outcry for reform. Sinclair was an ardent socialist who wrote his novel to expose and condemn the inequalities perpetuated by the avarice inherent in capitalism. In order to research his novel, Sinclair moved to Chicago in 1904 and, dressed as a worker, moved freely about the stockyards and slaughter

Figure 15.1 Dr. Harvey Washington Wiley (1844–1930) an analytical chemist and physician from Indiana whose lifelong passion for food and drug safety was rewarded in 1906 with the passage of the landmark Pure Food and Drug Act.

Source: Library of Congress, Prints & Photographs Division, Harvey W. Wiley Collection, LC-USZ62-96398.

houses to observe the conditions and to interview workers. Due to the radical content of his novel, he had a difficult time finding a publisher, but eventually did, and the novel sold by the millions. Despite Sinclair's radicalism, the novel struck a chord in the public consciousness, prompting President Theodore Roosevelt to order an investigation of the Chicago stockyards, resulting in reforms.[11]

By 1906 the Progressive Era was at its height and progressives had a strong advocate in the White House with President Theodore Roosevelt. With the advocacy of Dr. Harvey Wiley and his coalition of reformers, Congress passed the Pure Food and Drug Act on June 29, 1906 by a Senate vote of 63 to 4 in favor and a House of Representatives vote in favor of 241 to 17.[12] The law became the first federal legislation that dealt with drugs on the basis of adulteration, production, distribution, marketing, and import and export. The Act defined products that were to be regulated and defined acts of misconduct on the part of drug manufacturers. Section seven of the Act officially recognized the *U.S. Pharmacopoeia* and the *National Formulary* as works that defined drug standards. Section eight of the law defined a misbranded drug as one in which the labeling misrepresents the actual content of the drug. The law called for the mandatory labeling of the quantities of alcohol and narcotics including morphine, cocaine, or heroin that a nostrum contained. Most importantly, the law prohibited the interstate commerce of mislabeled drugs.[13] Dr. Wiley's U.S. Bureau of Chemistry (which later became the Food and Drug Administration (FDA) in 1931) was designated as the agency charged with the enforcement of the new law. As with many laws there were significant loopholes that allowed drug manufacturers to circumvent certain provisions, but reformers were optimistic that future legislation would close these loopholes. As a result of the new federal law nearly two-thirds of the states revised and updated their food and drug laws.

> In 1927, the precursor to the FDA was called the Food, Drug, and Insecticide Administration.

It did not take long for patent medicine makers to exploit the loopholes in the 1906 law. The law still allowed medicine makers to use narcotics and alcohol in their remedies as long as they were listed on the label. Some patent medicine makers proudly listed alcohol on their labels and specifically cited the 1906 law as an endorsement of the wholesomeness of their product! While the law regulated the labels on the medicine bottles, it did not regulate the outrageous therapeutic claims that were made in print advertisements, which were the lifeblood of the patent medicine industry. Ironically, the law did not regulate cosmetics, nor did it prohibit the use of poisons such as arsenic. The 1906 Act would be tested in the courts and in a split 1911 Supreme Court ruling, the Court ruled in *United States v. Johnson 221 U.S. 488* that the law did not prohibit false therapeutic claims, only the false labeling of ingredients.[14] Undaunted, Congress passed the Sherley Amendment in August 1912, which attempted to close the original loophole by prohibiting labels that made fraudulent claims about therapeutic benefits of a product. Unfortunately, the Sherley Amendment made it difficult for plaintiffs because they had to prove that the medicine manufacturer deliberately intended to make false claims, which was a very high legal burden of proof to achieve.[15]

The Harrison Narcotics Act of 1914

Since ancient times humanity has used opium and its derivatives for medicinal and later for recreational purposes. By the early twentieth century, several powerful and addictive narcotics had been produced through the wonders of alkaloid chemistry. Powerful opiates, including morphine and heroin, could be found in popular patent medicines. Similarly, cocaine, which was derived from the South American coca leaf, was even an ingredient in John Pemberton's original recipe for Coca-Cola. These relatively new and powerful narcotics showed great promise when used under controlled medical supervision, but soon became widely used for recreational purposes, which led to social problems associated with drug addiction. International leaders met and signed the Hague Treaty in 1912, in which the signatory nations pledged to cooperate to stop the international trafficking of narcotics. Prompted by this treaty as well as by domestic drug addiction problems, Congress passed the Harrison Narcotics Act on December 17, 1914, which regulated how narcotics would be distributed in the United States. The Act also taxed all narcotics that were produced or imported into the United States. Over time many states passed similar legislation.[16]

This Act was the first federal regulation of prescription drugs which changed the way pharmacists filled prescriptions. In order to dispense prescriptions that contained narcotics pharmacists had to register their stores with the Internal Revenue Service and pay a fee. Pharmacists could only dispense narcotics with a written prescription from an authorized physician or dentist. There were some exceptions made for topical preparations and for prescriptions containing a minimal amount of narcotics. Nonetheless, the Harrison Narcotic Act represented the first time the federal government placed restrictions on prescription drugs.[17]

Little did the legislators know what a monumental task awaited them in terms of drug enforcement. Over the next decades the federal government created the Federal Bureau of Narcotics in 1930, the precursor to the Drug Enforcement Administration, to enforce the laws and to disrupt the illegal trafficking of narcotics and harmful drugs. The Harrison Narcotics Act has been amended several times over the decades and one of the most important was the Drug Abuse Control Amendments of 1965. In response, the APhA published the *Record Book of Stimulant and Depressant Drug Transactions* to help pharmacists comply with mandatory inspections of controlled substance inventories. This book also included press-on labels that had to be placed on all controlled substances.[18] Since the 1970s, the United States has been engaged in an epic struggle to deal with the abuses of narcotic drugs and the social problems associated with their use.

The Trading With the Enemy Act of 1917

Although a few American pharmaceutical companies had developed research divisions to develop new drugs, World War I demonstrated that the majority of drugs Americans depended on were patented and controlled by foreign companies. Drugs such as aspirin, Salvarsan, and Novocain had been patented by German pharmaceutical companies. With the British Royal Naval blockade of Germany and its allies in force, neutral nations such as the United States could not receive the drugs, dyes, and chemicals they needed, even if the German companies were still willing to do business with American companies. In order to alleviate these shortages Congress took a bold step and passed the Trading With the Enemy Act of 1917 (TWEA) which abrogated drug patents held by German companies. In addition, the TWEA authorized the Federal Trade Commission (FTC) to issue licenses to American pharmaceutical companies to produce these drugs. The TWEA also put price controls for the drugs in place. Furthermore, the TWEA provided the FTC with strong pre-market regulatory authority over these drugs, which enhanced their safety.[19]

The Volstead Act of 1922 (Prohibition) 18th Amendment

Prohibition was both a blessing and a curse to most community pharmacists. As bars closed or became underground "speakeasies," community pharmacists who had soda fountains in their stores saw their sales soar with all the new traffic. By contrast, the Volstead Act made alcohol a prescription drug, which placed a heavy moral and psychological burden on pharmacists, especially those who tried to follow the new law conscientiously. The subsequent Willis-Campbell Act of 1921 exempted medicinal alcohol as long as it was prepared according to the standards of the *United States Pharmacopoeia* and the *National Formulary*. The law did restrict the amount of wine and spirits that could be prescribed. A physician was restricted to writing no more than 100 such prescriptions in 90 days. Patients could not receive more than a pint of medicinal alcohol once every ten days. Physicians had to use special prescription forms that were later redesigned by the Internal Revenue Service to thwart counterfeiting. The blank prescription forms also allowed for no refills. Physicians, pharmacists, and wholesalers had to obtain government permits to prescribe, dispense, or sell alcohol.[20]

Ironically, Prohibition also created a new demand for patent medicines, especially those that contained alcohol. The sale of alcohol-based patent medicines coupled with bitters became hot selling items during Prohibition. For example, Peruna, a patent medicine containing 18 percent alcohol, became especially popular due to a new medium, radio advertising.[21] One of the risks that had been associated with the sale of patent medicines was their safety. A popular patent medicine that was sold to circumvent the Volstead Act was "Tincture of Jamaica Ginger." Although it was a tincture, it could be imbibed directly from the bottle. To curb such abuse, the Prohibition Agency ruled that the nonprescription version of the drink could only contain fluid extract of ginger, which made it virtually impossible to drink. A Boston manufacturer, Hub Products Company, concocted a version of the drink using triorthocresyl phosphate, a common ingredient used in lacquers. By February 1930, the first reports emerged of people becoming paralyzed by imbibing this drink. Before it was all over, somewhere between 35,000 to 50,000 people had been killed or seriously disabled by this drink. The precursor to the FDA recalled and seized the products and prosecuted those involved in producing the adulterated substance. The prosecution of those involved in the harming of tens of thousands of people led to receiving fines of $1,000 with two years' probation and no prison time. FDA officials and prosecutors were stunned by the lenient sentences, revealing the weakness of 1906 Act, even with the 1912 Sherley Amendment. Although the sulfanilamide and thalidomide tragedies affected far fewer people, the Jamaica Ginger incident received virtually no media coverage, possibly because it affected adults who broke the law and lived mostly in poor rural areas.[22]

The Food, Drug, and Cosmetic Act of 1938

In 1933, the recently renamed Food and Drug Administration (FDA) engaged in a campaign to educate the public about food and drug safety with an agenda to amend and strengthen the 1906 Act. The FDA used educational displays called the "Chamber of Horrors" developed by FDA Chief Inspector George P. Harrick at 61 agricultural fairs nationwide to educate nearly 8 million people about food and drug safety. At the FDA display at the 1933 World's Fair in Chicago, visitors saw poster displays that demonstrated the loopholes in the 1906 Act. One display showed a woman who had been blinded by cosmetics. Another poster displayed the death certificates of diabetic patients who had once promoted a patent medicine cure for diabetes. Other displays showed examples of misleading packaging techniques, faulty medical devices, filthy factories, and contaminated foods. Ruth De Forest Lamb, an FDA official, wrote a book based on the exhibit entitled *The American Chamber of Horrors* that helped draw public attention for the need for food and drug safety.[23]

Despite reformers' efforts, it took another national tragedy to amend the 1906 Act. One of the earliest sulfa drugs to reach the market was sulfanilamide produced by the S.E. Masengill Company headquartered in Bristol, Tennessee. Sulfanilamide was used to treat streptococcal infections. The standard dosage form was a tablet or powder. Masengill's chief chemist, Harold Cole Watkins, wanted to produce the drug in liquid form and the solvent he chose was diethylene glycol, which is the key ingredient used today in automobile antifreeze. There were no legal requirements to test the product for toxicity, so the company tested for taste (raspberry), smell, and appearance. Masengill shipped its Elixir Sulfanilamide nationwide in September 1937. Within two months, over 100 people in 15 states, mostly children, had died from diethylene glycol poisoning. The first mass drug recall occurred and of the 240 gallons manufactured, 234 gallons and one pint were recovered through the diligent work of nearly all of the FDA's 239 agents.[24] In 1938, Samuel Evans Masengill was prosecuted for shipping a mislabeled drug, was convicted and fined $26,000. One of the pharmacists who had prepared the elixir later committed suicide, although he had not been charged with any crime.[25]

This tragedy prompted Congress to take action to protect the public from harmful foods, drugs, and cosmetics. Senator Burton Kendall Wheeler from Montana and Congressman Clarence F. Lea of California had been working on bills to amend the Federal Trade Commission Act of 1914 to provide the FTC with real power to ensure that manufacturers not engage in unfair or deceptive trade practices. President Franklin Delano Roosevelt signed the Wheeler-Lea Amendment on March 22, 1938, which empowered the FTC to prevent "any unfair method of competition or unfair or deceptive act or practice in commerce."[26] The law went into effect on May 21, 1938 and permitted the FTC to take action against false advertising and could levy fines from $500 to $5,000 depending on the severity of the infraction.[27]

Hero of FDCA: New York Senator Royal Copeland

Senator Royal Samuel Copeland of New York was the driving force behind the Federal Food, Drug, and Cosmetic Act of 1938. Copeland was born in Michigan in 1868 and as a young man worked for a homeopathic physician. Copeland went on to the University of Michigan where he earned his degree in homeopathic medicine in 1889 and practiced as an ophthalmologist and surgeon. He became mayor of Ann Arbor and served from 1901–1903.

In 1908, Copeland became dean of the New York Homeopathic Medical College where he led the Flower Hospital, a teaching hospital. He became New York City's commissioner of public health and served as president of the New York Board of Health until 1923. In 1924 he ran for the Senate and served until his death on June 17, 1938. Throughout his life Copeland was a tireless advocate for legislation that would protect consumers from harmful foods, drugs, and cosmetics, and lived to see it.[28]

President Franklin D. Roosevelt signed the Food, Drug, and Cosmetic Act of 1938 (FDCA), which gave the federal government sweeping power to control all medical devices and cosmetics. This law also had a new provision that called for a scientific review of all new medications before they could be sold on the market. This new drug application (NDA) has been modified in recent years to include the investigational new drug (IND) as well as clinical trials in animals and humans. The NDA objectives provided the FDA reviewer with data to determine if a proposed drug is safe and effective, that the benefits outweigh the risks, that the labeling is accurate, and the durability of the packaging will hold the strength, quality, and purity of the drug for its proposed shelf life.[29] The FDCA of 1938 was subsequently amended in 1941 to cover biological products such as insulin and, later, antibiotics. The Miller Amendment of 1946 extended the FDCA to cover products in transit in interstate commerce.[30]

The Durham–Humphrey Amendment of 1951

While the FDCA of 1938 represented landmark legislation that advanced drug safety greatly, two issues that it did not settle was what precisely constituted a prescription-only drug and what constituted a legitimate prescription

Figure 15.2 Carl Durham (1892–1974) an 11 term congressman and pharmacist from North Carolina, his name is forever enshrined in the landmark Durham–Humphrey Amendment (1951), which most notably clearly defined prescription and nonprescription products.

Source: Library of Congress, Prints & Photographs Division, Harris & Ewing Collection, LC-DIG-hec-26032.

order. This confusion led to *United States v. Sullivan* that reached the Supreme Court in 1948 in which the justices ruled that the pharmacist in question had violated federal law by selling a patient a restricted drug without a physician's prescription order. These issues were left to Congress to decide and define. Fortunately, there was a pharmacist in the House of Representatives named Carl T. Durham (1892–1974) from North Carolina who was uniquely qualified to remedy this confusion by introducing an amendment to the FDCA of 1938. Durham began introducing amendments in 1949, but several of his measures failed until he teamed up with Senator Hubert Humphrey from Minnesota who, similar to Durham, had worked in a pharmacy and was knowledgeable about the issues involved. On October 26, 1951 President Harry S. Truman signed the Durham–Humphrey amendment into law which was later implemented in April 1952. This law created two classes of drugs: prescription-only drugs that could be dispensed and purchased by order of a physician and nonprescription or over-the-counter drugs that had to be labeled but did not require a physician's prescription. The law also settled the nagging question about what constituted a legitimate prescription order by stating that telephone prescription orders and refill orders were legitimate. It also provided clear guidance to manufacturers about labeling requirements.[31]

Impact of the Durham–Humphrey Amendment on Pharmacy Practice

The post-war 1950s marked an era of unrivaled economic growth and prosperity for American pharmacy. For the first time in history pharmacists were dispensing wonder drugs such as antibiotics that rendered infectious bacterial diseases nearly obsolete curiosities of a distant past. Another first was that prescription business became the most profitable department in the drugstore for the first time in American history. According to the distinguished pharmacy historian, Gregory Higby, the era of "count and pour" pharmacy was born during this era of manufactured drugs and persists in some community pharmacies today. Higby argues that the Durham–Humphrey amendment, "restricted [pharmacists] to machine-like tasks" and "removed much of the pharmacist's autonomy in practice and the APhA Code of Ethics made the pharmacists' limited role quite clear."[32] Essentially, the APhA's 1952 Code of Ethics for pharmacists at that time prohibited pharmacists from discussing prescriptions with patients and referred them to discuss their medications with their prescriber. The APhA did not remove the clause about pharmacists not counseling patients about their drugs until 1969.[33] Higby adds, "Pharmacists gained respect from their connection with the new, effective drugs coming on the market, but their new reputation came at the cost of being considered overeducated for a diminished professional function."[34] In fact, prior to the FDCA of 1938 pharmacists could prescribe any non-narcotic drug to a patient. Durham–Humphrey and subsequent state anti-substitution laws transformed the pharmacist from a drug therapist into a product-dispenser.[35] Some pharmacists took comfort that they did not have any legal responsibility for the diagnosis and therapeutic outcome, only for the accurate processing and dispensing of the correct drug in the correct dosage, while others lamented and became frustrated with their more circumscribed role in health care. The latter group of pharmacists would spend the second half of the twentieth century trying to recover their role as key players in recommending drug therapies for patients.

The Evolution of State Anti-Substitution Laws from the 1950s to the Present

Boosted by the scientific breakthroughs made during and after World War II, the top 20 pharmaceutical manufacturers supplied about 75 percent of the medications (in dollar value) in the United States for newly written prescriptions by the late 1960s. In 1953, the National Pharmaceutical Council (NPC) was established to represent brand-name manufacturers. The NPC was devoted to lobbying for legislation that would prohibit pharmacists from substituting a brand-name drug for another product without the prescribing physician's authorization. By the early 1960s, the NPC's efforts in this area proved to be very successful in that most states passed legislation or created regulations which expressly prohibited the substitution of one drug for another without physician authorization.[36]

In 1970, the House of Delegates of the APhA took an official position which called for the repeal of anti-substitution laws. This position sparked considerable debate among APhA members, but in 1971, the APhA's House of Delegates took a stronger position, calling for assisting state pharmaceutical associations and legislatures to change the laws to allow pharmacists to select the supply of drugs they dispense. In 1972, the Commonwealth of Kentucky was the first to modify its anti-substitution law. By 1981, Indiana became the last state to change its anti-substitution laws, providing pharmacists with greater professional autonomy and judgment about patient drug therapies and complementing other efforts by the proponents of clinical pharmacy to transform the profession toward patient-centered care.[37]

The Kefauver–Harris Drug Amendments of 1962

During the baby-boom years of the 1950s a new drug called thalidomide came on the market in Canada and Europe (it was never approved for marketing in the United States) which was widely sold as a sedative to women to mitigate discomfort associated with pregnancy. Tragedy soon struck when increased numbers of infants with severely deformed limbs, a condition known as phocomelia, were born nationwide. An investigation traced the cause of the birth defects back to the use of thalidomide during pregnancy. In response to this tragedy, Congressional hearings were conducted for three years, beginning in December 1959, led by Senator Estes Kefauver of Tennessee. In 1962, the Kefauver–Harris amendments were passed and ushered in a new era of FDA oversight of drug safety. The law placed the burden of drug safety squarely on the shoulders of drug manufacturers by increasing FDA authority over patient consent, clinical drug trials involving humans, drug factory inspections, and oversight over prescription drug advertising. The law did not grandfather in existing drugs, so drug makers had to prove the safety of all prescription drugs made from 1938 to 1952. In 1966, the FDA appointed panels of experts from the National Academy of Science and the National Research Council to evaluate all of these drugs. As a result of this evaluation, over 7,000 prescription drugs were taken off the market, with another 1,500 relabeled. In 1972, a similar evaluation was launched for nonprescription drugs. The Kefauver–Harris amendments provided more drug safety, but increased the time and cost of bringing new drugs to market, thus driving up the prices of prescription drugs.[38]

Dr. Frances Oldham Kelsey was a pharmacologist who hailed from Vancouver, Canada and became an American citizen in 1955. She moved to Washington, DC in 1960 when her husband became a special assistant to Surgeon General Luther L. Terry. She became a medical officer with the FDA in August 1960. Her first task was to review thalidomide, a new drug being offered by William S. Merrell Company. She refused to authorize the company's application to approve the drug, on the grounds that it might lead to peripheral neuritis. In 1962, the FDA Commissioner George P. Larrick announced that some American women had been given thalidomide and that William S. Merrell Company had distributed the drug to 1,200 American physicians. Sadly, 17 American children suffered from severe birth defects as a result of their mothers having taken the drug.[39]

In August 1962, Dr. Kelsey testified before the government operations subcommittee that she had suspected thalidomide of causing peripheral neuritis in February 1961 and notified the manufacturer. She suspected that the drug might have an adverse effect on human fetuses in May 1961. In West Germany, where the drug first originated, the government withdrew thalidomide from the market because of birth defects. Nonetheless, the Merrell Company continued to lobby for the drug's approval. Kelsey, with the backing of her supervisors at the FDA, refused to approve the drug.

Shortly after her testimony, Kelsey became a national hero and a minor media sensation for her unusual resourcefulness and courage. On August 7, 1962, Dr. Kelsey received the President's Award for Distinguished Service from President John F. Kennedy. Dr. Kelsey's work and example were keys to gaining the public support needed to pass the Kefauver–Harris Amendment.

Figure 15.3 Dr. Frances O. Kelsey (1914–2015) is a Canadian-born pharmacologist, who conducted an investigation on behalf of the FDA about the potential harmful effects of thalidomide, and refused to allow the drug to go on the market. In August 1962, President JFK bestowed the President's Distinguished Service Award on her. In 2010, Dr. Kelsey was the first recipient of the Dr. Francis O. Kelsey Award for Excellence and Courage in Protecting Public Health.

Source: Library of Congress, Prints & Photographs Division, New York World-Telegram and the Sun Newspaper Photograph Collection, LC-USZ62-131536.

Kelsey went on to a distinguished career in the FDA and was named director of the Investigational branch in 1962, and later in 1981 she headed the Division of Scientific Investigations in the Bureau of Drugs. Her staff became known as "Kelsey's Kops."[40] She continued to serve the FDA as well as the American public well into her nineties.

Title XVIII Social Security Act of 1965 (Medicare Parts A & B) and Title XIX Social Security Act of 1965 (Medicaid)

Unlike the aforementioned federal laws, the introduction of Medicare and Medicaid were not designed to improve drug safety per se, but to expand access to health care and demand for pharmacy services for senior citizens and the indigent. These laws put the federal government into the American health care system as a major player and affected pharmacy practice dramatically by becoming the largest third-party payer for prescription drugs today. For example, in 1968 Medicare and Medicaid accounted for 10 percent of all payments for prescription drugs. By 1993, third-party payments (both government and private insurance) accounted for nearly half of all such payments. In the wake of Medicare Part D and the expansion of private health insurance, the 1968 scenario had been reversed completely by 2008, with third parties paying for 90 percent of all prescription drugs, with the bulk of that coming from the government.[41] In 1969, the Department of Health, Education, and Welfare (now the Department of Health and Human Services) created a task force to study the possibility of adding a prescription drug component to Medicare and studied the idea of having pharmacists play a more active role in counseling patients about their medications.[42]

The Poison Prevention Packaging Act 1970

Modern drugs represent modern miracles and have been a great benefit to society, but their power posed great hazards to young children who might ingest them accidentally. During the 1960s accidental poisoning became the leading cause of death for children under the age of five. Prompted by the problem and the results of a study conducted at the Fort Lewis–McChord Air Force military bases in Tacoma, Washington, child-resistant prescription bottle caps were invented. In a pilot study, military families were given regular prescription bottles with screw lids and new push-down-and-turn lids. The study showed the effectiveness of the new lids and a similar study conducted in Ontario, Canada, yielded similar positive results in keeping drugs out of the hands of young children. In 1970, Congress passed the Poison Prevention Packaging Act which required child-resistant packaging for certain over-the-counter medications and for prescription drugs as well. To accommodate individuals with special needs, patients can request non-child-resistant lids.[43]

The 1990 Omnibus Budget Reconciliation Act

Since the Durham–Humphrey amendments of 1952, the pharmacist's role was restricted largely to dispensing drugs accurately and sometimes this resulted in tragic results. Such was the case of the owner of T.I. Drugstore, Henry Fan, and one of his patients, Elaine Jones. In what at first appeared to be routine, one of Fan's pharmacists filled a prescription for Benedictin for Jones. Eight months later Jones gave birth to a baby girl with a severe spinal birth defect. Jones sued T.I. Drugstores for failing to counsel her about the possible side effects of the medicine. The attorney for the drugstore argued that it was the prescribing physician's responsibility under the law to counsel a patient about the medication and its possible side effects. Furthermore, the attorney representing the drugstore argued that his client had a duty to dispense the drug correctly as prescribed, which he did.

On November 5, 1990, as part of an overall budget bill President George H.W. Bush signed the Omnibus Budget Reconciliation Act (OBRA 1990) into law. Within a few years, with very rare exceptions, most states enacted similar laws. The Act included several provisions about Medicaid that affected the way pharmacy in general would be practiced. Taking effect in 1993, OBRA mandated a drug-utilization review (DUR) for all Medicaid patients. A DUR calls for the evaluation of a drug for its cost and medical necessity. Most importantly, OBRA required pharmacists to offer counseling to their patients about the possible side effects and interactions with other medications. Although OBRA only applied to Medicaid patients, seeing its utility for all patients, nearly all states mandated its use for all patients. OBRA also called for studies to examine cost-benefit analysis of DUR and the impact of paying pharmacists for counseling or cognitive services. OBRA caused pharmacists to shift their focus from dispensing drug products to patient-centered care.

OBRA triggered a few important changes in how pharmacists came to view themselves and the purposes of their practices. In 1991, the Joint Commission of Pharmacy Practitioners (JCPP) drafted a new mission statement that was accepted by a majority of national pharmacy organizations that promoted the idea of helping patients get the most from their medicines.[44] Looking toward fulfilling the spirit of OBRA, the APhA revised the Pharmacists' Code of Ethics in 1994 to reflect an emphasis on assuming professional responsibility for optimal patient drug therapy outcomes. A far cry from the 1952 call to simply get the prescription correct, the 1994 code called upon pharmacists to put their patients' health outcomes before all else and to assume personal responsibility for them.[45]

The Medicare Prescription Drug Improvement and Modernization Act 2003

One of the components of the Medicare program that was missing was a prescription drug benefit. After decades of debate, the most significant expansion of the Medicare program occurred when Congress passed the Medicare Prescription Drug Improvement and Modernization Act in October 2003. Scheduled to take effect in 2006, Medicare Part D as it came to be called, changed pharmacy practice significantly with the implementation of medication therapy management (MTM) services for seniors. MTM became an important corollary to pharmaceutical care in that it was designed for pharmacists to monitor a patient's drug therapy to yield optimal health outcomes.

Chapter Summary

From the time the United States Constitution was written in 1787, law has profoundly affected the way pharmacy has been practiced. The first patent laws encouraged and protected drug makers which ushered in the proliferation of unregulated patent medicines. Guided by laissez-faire economic principles, patent medicines flourished under a system that provided few legal safeguards for the public. Though the Drug Importation Act of 1848 was designed to protect the public from faulty imported drugs, it was not until the Pure Food and Drug Act was passed in 1906 that the American public was afforded some protection against harmful drugs. The Food, Drug, and Cosmetic Act of 1938 closed several loopholes in the 1906 law and extended regulation over cosmetics. The Durham–Humphrey Amendment in 1951 created two distinct classes of drugs: prescription and nonprescription and the Kefauver–Harris Amendments in 1962 strengthened previous legislation and gave the FDA greater powers of oversight in ensuring a safe drug supply.

After the Kefauver-Harris Amendments ensured greater drug safety, the federal government embarked upon a new era of legislation that would extend access to health care to senior citizens and the indigent with the passage of legislation that created the Medicare and Medicaid programs in 1965. This landmark legislation made the federal government a major player in the health care system and also made it a catalyst for change in the way health care was delivered. Medicare and Medicaid prompted several important changes in the way pharmacy was practiced in the second half of the twentieth century and beyond. For example, in 1968 when Medicare and Medicaid took effect, third-party payers for prescription drugs made up only 10 percent of all drug sales in the United States. By 2008, the U.S. government was the largest third-party payer for all prescription drugs in the US, and coupled with private third-party payer insurance companies accounted for 90 percent of all payments for prescription drugs. One of the great failures of the past century is that dietary supplements remain virtually unregulated and leave the public in much the same position it was back in 1900, when desperate people searched for miracle cures.

The OBRA of 1990 introduced mandatory drug-utilization review (DUR) for all Medicaid patients, which was extended later to include all patients. Similarly, the Medicare Prescription Drug Improvement and Modernization Act of 2003 introduced MTM for all Medicare patients. At the start of the twentieth century, pharmacy practice was largely unregulated and with the Humphrey–Durham amendment in 1951 became much more circumscribed. During the second half of the twentieth century and beyond, pharmacists have embarked upon a quest to regain much of their professional autonomy as vital partners of the modern health care team as drug therapy experts.

Key Terms

Patent medicines

Drug Importation Act of 1848

Biologics Control Act of 1902

The Pure Food and Drug Act of 1906

Harvey Washington Wiley

Upton Sinclair's *The Jungle*

US Bureau of Chemistry

Food and Drug Administration (FDA)

United States v. Johnson

Sherley Amendment 1912

Harrison Narcotics Act of 1914

Federal Bureau of Narcotics

Drug Enforcement Agency

Trading With the Enemy Act of 1917

The Volstead Act of 1922 (Prohibition) 18th Amendment

Royal Copeland

The Food, Drug, and Cosmetic Act of 1938

Sulfanilamide scandal

Durham–Humphrey Amendment of 1951

Era of "Count and Pour"

Kefauver–Harris Amendments of 1962

Thalidomide scandal

Frances Oldham Kelsey

Title XVIII Social Security Act of 1965 (Medicare Parts A & B)

Title XIX Social Security Act of 1965 (Medicaid)

Poison Prevention Packaging Act 1970

The 1990 Omnibus Budget Reconciliation Act (OBRA 1990)

drug-utilization review (DUR)

medication therapy management (MTM)

Medicare Prescription Drug Improvement and Modernization Act 2003

Chapter in Review

1 Define what a patent medicine is and account for their popularity in American culture.

2 Describe how the Drug Importation Act of 1848 came about and its historical significance.

3 Describe the conditions that led to the passage of the Pure Food and Drug Act of 1906 and its impact on future drug safety legislation.

4 Explain the impact of the Volstead Act (Prohibition) on the practice of pharmacy.

5 Explain the historical significance of the Food, Drug, and Cosmetic Act of 1938.

6 Describe the impact the Durham–Humphrey Amendment had on the practice of pharmacy.
7 Describe the impact the Kefauver–Harris Amendment had on drug safety.
8 Explain how Medicare and Medicaid have impacted the practice of pharmacy.
9 Explain the relationship of federal legislation and the practice of pharmacy.
10 Explain how the OBRA 1990 Act changed the way pharmacy was practiced.
11 Explain the significance of the Medicare Prescription Drug Improvement and Modernization Act of 2003.

Notes

1 David L. Cowen and William H. Helfand, *Pharmacy: An Illustrated History* (New York: Harry Abrams, 1990), 167.
2 David Armstrong and Elizabeth Metzger Armstrong, *The Great American Medicine Show* (New York: Prentice-Hall, 1991), 159.
3 George Griffenhagen, *150 Years of Caring: A Pictorial History of the American Pharmaceutical Association* (Washington, DC: American Pharmaceutical Association, 2002), 129.
4 W. Stephen Pray, *A History of Nonprescription Drug Regulation* (New York: Pharmaceutical Products Press, 2003), 6.
5 Gregory J. Higby, "From Compounding to Caring: An Abridged History of American Pharmacy." In *Pharmaceutical Care*, edited by Calvin H. Knowlton and Richard P. Penna (Bethesda, MD: American Society of Health-System Pharmacists, 2003), 28.
6 W. Stephen Pray, *A History of Nonprescription Drug Regulation*, 6.
7 Ramunas A. Kondratus, "The Biologics Control Act of 1902." In Glenn Sonnedecker, *The Early Years of Food and Drug Control* (Madison, WI: AIHP, 1982), 14. Glenn Sonnedecker, comp., *Kremers and Urdang's History of Pharmacy* (Madison, WI: AIHP, 1976), 221. Dennis Worthen, "The Pharmaceutical Industry, 1902–1952." In *American Pharmacy: A Collection of Historical Essays*, edited by Gregory Higby (Madison, WI: AIHP, 2005), 61–62.
8 David Armstrong and Elizabeth Metzger Armstrong, *The Great American Medicine Show*, 169. W. Stephen Pray, *A History of Nonprescription Drug Regulation*, 7–8.
9 Ibid. (Pray), 58–59. Charles O. Jackson, *Food and Drug Legislation in the New Deal* (Princeton, NJ: Princeton University Press, 1970), 4.
10 David Armstrong and Elizabeth Metzger Armstrong, *The Great American Medicine Show*, 169.
11 W. Stephen Pray, *A History of Nonprescription Drug Regulation*, 35–36.
12 Ilyse D. Barkan, "Industry Invites Regulation: The Passage of the Pure Food and Drug Act of 1906," *American Journal of Public Health*, Volume 75 (January 1985), Issue 1.
13 George Griffenhagen, *150 Years of Caring*, 131.
14 Ibid., 132.
15 W. Stephen Pray, *A History of Nonprescription Drug Regulation*, 53.
16 Ibid., 82–90. Joseph F. Spillane, "The Road to the Harrison Narcotics Act: Drugs and Their Control, 1875–1918." In *Federal Drug Control: The Evolution of Policy and Practice*,

edited by Jonathon Erlen and Joseph F. Spillane (New York: Pharmaceutical Products Press, 2004), 2–24.

17 John P. Swann, "FDA and the Practice of Pharmacy: Prescription Drug Regulation before the Durham-Humphrey Amendment of 1951." In *Pharmacy in History*, edited by Gregory Higby (Madison, WI: AIHP), Vol. 36 (1994), No. 2, 57.

18 George Griffenhagen, *150 Years of Caring*, 133.

19 Dale Cooper, "The Licensing of German Drug Patents Confiscated During World War I: Federal and Private Efforts to Maintain Control, Promote Production, and Protect Public Health." In *Pharmacy in History*, edited by Gregory Higby (Madison, WI: AIHP), Vol. 54 (2012), No. 1, 3.

20 John P. Swann, "FDA and the Practice of Pharmacy," 57.

21 David Armstrong and Elizabeth Metzger Armstrong, *The Great American Medicine Show*, 170.

22 W. Stephen Pray, *A History of Nonprescription Drug Regulation*, 56–57.

23 Ibid., 95–98. Gwen Kay, "Healthy Public Relations: The FDA's 1930s Legislative Campaign," *Bulletin of the History of Medicine*, Vol. 75 (2001), Issue 3, 446–487.

24 W. Stephen Pray, *A History of Nonprescription Drug Regulation*, 115. Carol Ballantine, "Taste of Raspberries, Taste of Death: The 1937 Elixir Sulfanilamide Incident," *FDA Consumer Magazine*, June 1981.

25 Ibid. (Pray), 118.

26 Ibid., 121.

27 Ibid., 121.

28 Ibid., 94.

29 FDA, "New Drug Application," 20 August 2010. http://www.fda.gov/Drugs/DevelopmentApprovalProcess/HowDrugsareDevelopedandApproved/Approval Applications/NewDrugApplicationNDA/default.htm.

30 George Griffenhagen, *150 Years of Caring*, 133.

31 Ibid., 134. Dennis B. Worthen, *Heroes of Pharmacy: Professional Leadership in Times of Change* (Washington, DC: American Pharmacists Association, 2008), 84. Dennis B. Worthen, "The Pharmaceutical Industry, 1952–2002." In *American Pharmacy: a Collection of Historical Essays*, edited by Gregory Higby (Madison, WI: AIHP, 2005), 70. W. Stephen Pray, *A History of Nonprescription Drugs*, 145.

32 Gregory Higby, "From Compounding to Caring: An Abridged History of American Pharmacy", in *Pharmaceutical Care*, Calvin H. Knowlton and Richard P. Penna, eds. (Bethesda, MD: American Society of Health-System Pharmacists, 2003), 37.

33 Robert Elenbaas and Dennis B. Worthen, eds., *Clinical Pharmacy in the United States: Transformation of a Profession* (Lenexa, KS: American College of Clinical Pharmacy, 2009), 38.

34 Gregory Higby, "From Compounding to Caring," 37.

35 David Brushwood, "Governance of Pharmacy, 1902–1952." In *American Pharmacy: A Collection of Historical Essays*, edited by Gregory Higby (Madison, WI: AIHP, 2005), 80.

36 Joseph L. Fink III, "Pharmacy and Public Policy: Evolution of the Legal and Regulatory Framework for Generic Interchange," *Journal of Pharmacy Technology*, Vol. 24 (May/June 2008), 131–133.

37 Ibid., 132.

38 George Griffenhagen, *150 Years of Caring*, 135.

39 W. Stephen Pray, *A History of Nonprescription Drugs*, 161.

40 Ibid., 170.

41 Stephen W. Schondelmeyer, "Recent Trends in American Pharmacy." In *Pharmacy in History*, edited by Gregory Higby (Madison, WI: AIHP) Vol. 51 (2009), No. 3, 113.

42 Robert M. Elenbaas and Dennis B. Worthen, eds., *Clinical Pharmacy in the United States: Transformation of a Profession* (Lenexa, KS: ACCP, 2009), 38.

43 Joseph L. Fink III, "Child-Resistant Packaging for Medication," *Drug Intelligence and Clinical Pharmacy*, 10, 1976(Dec), 698–702.

44 *American Pharmacist*, 1991; NS31, 711–712.

45 *American Pharmacist*, 1994; 34: 79.

16 Hospital Pharmacy and the Rise of Clinical Pharmacy in the United States

What factors contributed to the origin of clinical pharmacy and how did it transform the profession?

Although American hospital pharmacy dates back to Jonathan Roberts, John Morgan, and the founding of Pennsylvania Hospital in Philadelphia in 1751, modern hospitals, and subsequently clinical pharmacy, began in the 1920s (see Chapter 9, "Colonial and Early American Pharmacy"). In fact, Roberts and especially Morgan were emblematic of the fact that medical apprentices back then were in charge of hospital pharmacies, not apothecaries. This trend did change during the Early Republic when New York Hospital tested and hired an apothecary in 1811 to compound its prescriptions and manage its dispensary. By 1819 apothecaries, with ability to compound prescriptions and manage their dispensaries, had displaced medical apprentices who were being trained to write prescriptions rather than compound them.[1] Despite these early advances hospital pharmacists remained a forgotten minority, operating in the shadows of community pharmacists, even though hospital pharmacists continued to manufacture and compound drugs and special products after these practices were no longer performed in community pharmacies. Scientific advancements and the rise of the modern hospital changed all that. Clinical pharmacy was born coincidentally with the rise of the modern American hospital. By the 1920s hospitals evolved from passive warehouses for the sick and places to die, into places that actively held out hope for cures. With the rise of antibiotics and antiseptic surgery hospitals became places where medical miracles occurred.[2] During the 1920s, there were 6,000 hospitals employing about 500 pharmacists.[3]

The Earliest Calls for Clinical Pharmacy

1921 was a momentous year for both hospital and clinical pharmacy. In a seminal paper, E.C. Austin, a hospital pharmacist from Cincinnati, lamented the lack of well-trained hospital pharmacists. He also expressed the long-standing frustration of hospital pharmacists who had been marginalized in the profession and often treated as second class citizens at the hands of their more numerous community pharmacy colleagues.[4] At the annual meeting of the American Pharmaceutical Association (APhA) that year, John C. Krantz, Jr. from the University of Maryland proposed the idea of hospital pharmacists providing clinical services at the patient's bedside in addition to performing their traditional role of dispensing medicines. His vision included having pharmacists work in the laboratory conducting pathology tests and chemical analyses. While working in a hospital laboratory has not become part of what clinical pharmacy is today, Krantz's vision of getting the pharmacist out of the hospital pharmacy and dealing directly with patients and other health care professionals has come to fruition.[5]

At that same APhA meeting, the organization's president, Charles H. Packard, suggested that hospital pharmacists form a committee that would play an annual role in the Section on Practical Pharmacy and Dispensing. Since all of the officers in the Section in 1922 were hospital pharmacists, they proposed that a section of the APhA be created exclusively for them. Edward Swallow of Bellevue Hospital in New York wrote an article in the *APhA Journal* calling for hospital pharmacists to have their own section within the APhA. Swallow's article about hospital pharmacy was followed by others, which were featured regularly by the *APhA Journal*.[6]

One of the great pioneers and innovators of hospital pharmacy was Harvey A.K. Whitney (1894–1957). Whitney was born in Adrian, Michigan in 1894 and in 1913 got his first job in pharmacy as an apprentice in his hometown. After serving in the medical corps in World War I, he graduated from the University of Michigan's College of Pharmacy with his degree as a pharmaceutical chemist. In 1925, Whitney got a job at the University Hospital in Ann Arbor and in 1927 became chief pharmacist. Seeing a great need, Whitney introduced the first formal hospital internship program in the US in 1927, which became a model for today's residency programs. Over the next decade, Whitney developed a formulary, published pharmacy bulletins for the hospital staff, and founded a drug information center in the hospital. Whitney manufactured prepackaged medications for the hospital that were coded and dated for easy and safe use—a precursor to the unit dose. He also pioneered the manufacturing of sterile products for hospital use. These innovations came to define what a modern hospital pharmacy was all about.[7]

While Whitney was advancing hospital pharmacy in Michigan, Louis C. Zopf was making his own innovations at the University of Iowa Hospital. In 1928, Zopf introduced the practice of getting hospital pharmacists out of the hospital pharmacy and having them go on patient rounds with physicians as they tended to their patients. He also had the hospital pharmacists deliver seminars on drug therapies and other pharmacy related topics for the hospital staff.

Similarly, in 1932, Edward Spease, who became dean of the pharmacy school at Western Reserve University, forged an agreement with the

University Hospital to provide joint appointments for all pharmacists. Soon pharmacists were going on rounds with physicians, visiting patients, and had gained the support and confidence of physicians, most importantly Dr. Henry J. Goeckel. As a result of this experience, Spease presented a seminal paper at the 1935 meeting of the American College of Surgeons Clinical Congress. In his paper, Spease recommended that every hospital have a pharmacy staffed by a registered pharmacist who would be charged with obtaining and administering all medications, establish a Pharmacy and Therapeutics Committee, and establish a reference library. The American College of Surgeons (now known as the Joint Commission) adopted Spease's recommendations and made them standards. In 1937, Spease created a hospital internship program at Western Reserve University.[8]

In 1925, hospital pharmacists in California formed the first state association—the Hospital Pharmacy Association of Southern California. By 1939, California, Illinois, Indiana, Iowa, Minnesota, Nebraska, New York, Ohio, Pennsylvania, and Wisconsin all had state hospital pharmacy associations.[9] With the pressure building for greater recognition at the national level, hospital pharmacy leaders convinced the APhA leadership to organize a sub-section on hospital pharmacy under the Section on Practical Pharmacy and Dispensing in 1936. That same year the American Hospital Association (AHA) organized a Pharmacy and Therapeutics Committee with a pharmacist serving as a member. In 1937, the AHA's Committee on Pharmacy issued a report calling for patient safety guidelines which stated that any hospital with a hundred beds or more should require a registered pharmacist on the hospital's staff.[10] As early as 1940 there was a proposal to form an affiliate organization for hospital pharmacists with the APhA. On August 21, 1942, the American Society of Hospital Pharmacists (ASHP) was born in Denver as affiliate of the APhA. Members had to belong to APhA before they could belong to the new ASHP. Fittingly, Harvey A.K. Whitney served as the first president of ASHP at its first meeting in 1943. The ubiquitous Whitney helped establish the *Official Bulletin of the American Society of Hospital Pharmacists*, which he and Leo W. Mossman co-edited until Donald E. Francke and Gloria Niemeyer assumed editorship in 1944. The journal changed names several times over the years assuming its current moniker, the *American Journal of Health-System Pharmacy*, in 1995. Whitney also made a lasting contribution to hospital pharmacy by offering the formulary he had developed for his hospital back in the 1930s, which became the basis for the *American Hospital Formulary Service* published in 1959.[11]

With the rise of powerful new "wonder" drugs led by antibiotics, hospital pharmacists became major players in the golden age of the American hospital during the post-war years or baby-boom era. Pharmacists became key players in developing hospital formularies, which played increasingly important therapeutic roles in patient well-being. During this momentous time two pharmacists who were visionaries helped transform hospital pharmacy into modern clinical pharmacy: Donald Eugene Francke and Donald Crum Brodie.

Donald Eugene Francke (1910–1978) was born in Athens, Pennsylvania, and learned about pharmacy by working in his father's drugstore. Francke attended the University of Michigan, earning his B.S. in pharmacy in 1936 and worked for Harvey A.K. Whitney until going to Purdue University where he earned a master's degree in pharmaceutical chemistry in 1938. Francke returned to the University of Michigan Medical Center where he served as pharmacy director from 1944–1963, succeeding Whitney. In addition to his duties as pharmacy director tending to the hospital formulary and serving on the Pharmacy and Therapeutics Committee, Francke served as the director of the APhA's Division of Hospital Pharmacy from 1949–1955. Meanwhile, Francke directed a nationwide survey project that studied the various services hospital pharmacies provided and how they operated for the U.S. Public Health Service. The survey published its results and recommendations in a 1964 report entitled the *Mirror to Hospital Pharmacy*. This report provided a blueprint for transforming hospital pharmacy into clinical pharmacy and shed light on the roles pharmacists would play in the future.[12]

In addition to his long-standing editorship of hospital pharmacy's premier journal, which he edited until 1966, Francke pioneered the effort to provide hospital pharmacists with reliable drug information. The first effort was the *American Hospital Formulary Service* published in 1959. Today it is called the *American Hospital Formulary Service Drug Information* and profiles about 40,000 medications and is also available online.[13] Francke published *International Pharmacy Abstracts* in 1964 and served as editor of both until 1966, the same year that he became the founding editor of the *Drug Information Bulletin*. These efforts culminated in the journal called *Drug Intelligence and*

The First Drug Information Center

The first drug information center in the United States staffed by pharmacists was established in August 1962 at the University of Kentucky. The center came about as a result of discussions held in 1960 by Dr. Edward Pellegrino, who was chairman of the Department of Medicine, and Dr. Charles Walton, a faculty member of the University of Kentucky College of Pharmacy, who both saw the value of such a center to promote the use of rational drug therapy. Dr. David Burkholder became the Assistant Director of the Center which became a place where pharmacists played a key role in patient-centered drug therapy, which advanced the cause of clinical pharmacy.[14]

Clinical Pharmacy which premiered in 1969.[15] These editorial efforts provided the theory and ideas that guided the early development of clinical pharmacy in the United States. In addition to his editorial work Francke helped establish the first academic department devoted to the study of clinical pharmacy at the University of Cincinnati in 1968, which served as a national model.[16]

Another key figure in the development of clinical pharmacy was Donald Crum Brodie (1908–1994) who coined and defined the term *pharmaceutical care* in 1973. This is not surprising since, as early as 1953, Brodie spoke about combining the traditional services hospital pharmacists provided, but gearing them toward patient care. Similar to Francke, Brodie devoted his career to getting his pharmacy colleagues to shift their traditional focus on drugs toward patient care. Brodie was born in Carroll, Iowa, and worked in his father's drugstore. He attended the University of Southern California where he earned his bachelor's and master's degrees in 1934 and 1938, respectively. He earned his doctorate in pharmaceutical chemistry from Purdue University in 1944. Brodie joined the faculty at the University of California–San Francisco in 1947 and later became director of pharmaceutical services at the Moffitt Hospital at the UCSF Medical Center. Brodie, highly influenced by the work of Edward Elliott and the Pharmaceutical Survey's call for a transformation of pharmacy education and practice, endorsed the introduction of a six-year Pharm.D. program at UCSF in 1955. In 1966, Brodie and his associates, William E. Smith, Jr., Sidney Riegelman, Eric Owyang, Donald Sorby, and Jere Goyan, embarked on a bold experiment that came to be known as the Ninth Floor Pilot Project that transformed clinical pharmacy. A 24/7 satellite pharmacy was placed on a general surgery floor next to a nursing station to provide unit-dose drug distribution and parenteral services. The pharmacy provided consulting services, unit doses, pharmacy technicians, patient drug profiles, and a drug information center. Pharmacists took detailed patient medication histories upon patient admittance and provided medication counseling for patients before they were released. The Ninth Floor Project's success had profound implications for pharmacy education and practice and clearly demonstrated the value of clinical pharmacy to the health care community.

Brodie continued to be an advocate of drug-use control in order to get pharmacists to embrace a new role centered on taking professional responsibility for their patients' health outcomes. In 1967, Brodie promoted the idea of the pharmacist as an integral part of the health care team, arguing that pharmacy education needed to become more inter-professional and include more components of direct patient contact. In one of his papers, Brodie cited the pioneering efforts of Eugene V. White (1924–2011) who, since 1960, had compiled and maintained patient medication profiles in his community pharmacy. White's work in Berryville, Virginia, exemplified how clinical pharmacy could be practiced successfully in a community practice setting.[17] By 1969, clinical pharmacy had come to be defined by ASHP and AACP, and

Figure 16.1 Donald Crum Brodie (1908–1994) pioneered the concept of pharmaceutical care, i.e. shifting the pharmacist's focus from the product to the patient, a key feature of clinical pharmacy.

Source: Courtesy of the National Library of Medicine.

by 1980 the ranks of institutional pharmacists had grown to 40,000 representing about 25 percent of all pharmacy practitioners.[18] In 1969, a momentous change occurred when the APhA's Code of Ethics was revised and rescinded the long-standing recommendation that pharmacists not discuss drug therapy or drug action with patients. In fact, the 1969 revision stated, "A pharmacist should hold the health and safety of patients to be of first consideration; he should render to each patient the full measure of his ability as an essential health practitioner."[19]

In 1971, the National Center for Health Services Research and Development met to further define the role of clinical pharmacists. The ubiquitous Donald Brodie served on this task force and helped the group draft a detailed description of what clinical pharmacists could do to enhance optimal health outcomes for patients by providing patient-centered care in various practice settings. In the same year, a new docent clinical pharmacist program was established at the University of Missouri–Kansas City Medical School by none other than David Burkholder who helped establish the Drug Information Center at the University of Kentucky. From the start of this medical school, clinical pharmacy was deliberately integrated into the medical and pharmacy schools' curricula and established the clinical pharmacist as part of the health care team.[20] By 1972, clinical pharmacy and residency programs had reached a critical mass from Pharm.D. programs at northern and southern California, SUNY Buffalo, Illinois, Kansas City, Kentucky, Minnesota, Philadelphia, South Carolina, and Texas. The graduates of these programs provided the faculty to educate the next wave of clinical pharmacists.

The Millis Commission's report, *Pharmacists for the Future*, published in 1975, affirmed the need for clinical pharmacy and that patient care was the future of the profession. In the same year, clinical faculty members from UCSF and Washington State University collaborated on a book entitled *Applied Therapeutics for Clinical Pharmacists*. This important book was edited by Mary Anne Koda-Kimble, Lloyd Young, and Brian Katcher and created the blueprint for teaching therapeutics. Since 1983, the book's title has been *Applied Therapeutics: The Clinical Use of Drugs*. Another major addition to the growing body of literature was *Pharmacotherapy: A Pathophysiologic Approach*, edited by Joseph DiPiro, Robert Talbot, Peggy Hayes, Gary Yee, and L. Michael Posey. The first edition appeared in 1988 and the wide-ranging text has gone through several editions and is often referred to simply as "DiPiro."[21] In 1979, clinical pharmacists had their own organization—the American College of Clinical Pharmacy (ACCP). ACCP was founded as the result of a meeting in September 1979 at the University of Missouri–Kansas City. With 29 founding members, Kim Kelly was elected as its first president. Beginning in 1982, ACCP worked diligently to get clinical pharmacy recognized as a board specialty, but the Board of Pharmaceutical Specialties (BPS) decided not to recognize clinical pharmacy as a specialty on the grounds that it was too broad in nature. Undaunted, ACCP submitted a petition to get pharmacotherapy recognized as a specialty and their petition was approved in 1988, with the first certification examinations administered in 1991.

Table 16.1 Specialties Recognized

Nuclear Pharmacy	1978
Pharmacotherapy	1988
Parenteral and Enteral Nutrition	1988
Psychiatry	1992
Oncology	1996
Sub-Specialties	
Infectious Disease Pharmacotherapy	1999
Cardiology Pharmacotherapy	2000
Ambulatory Care	2008

At about the time ACCP was getting organized and some clinical pharmacists were serving as principal investigators in research projects, a second generation of literature concerning clinical pharmacy emerged in the 1980s. In 1981, Russell Miller founded a new journal called *Pharmacotherapy* that would feature some of the pharmacology research being conducted by clinical pharmacists. The journal emerged out of the backdrop of the Boston Collaborative Drug Surveillance Program conducted at the Boston University Medical Center. *Pharmacotherapy* became ACCP's journal in 1988. Similarly, another important journal, *Clinical Pharmacy*, was founded in 1982 by ASHP and was edited by William Zellmer. The journal devoted itself to publishing novel therapeutic drug therapies for a broad audience of pharmacy practitioners. In 1994, this journal became part of the *American Journal of Hospital Pharmacy*.[22]

Pharmaceutical Care

By the 1980s, clinical pharmacy took on the aspects of a social movement within the pharmacy profession that called for its members to shift their focus from dispensing medications to caring for patients regardless of their practice setting. In 1985, ASHP held an invitational conference at Hilton Head, South Carolina to define and advance clinical pharmacy. One of the keynote speakers was C. Douglas Hepler who delivered a speech, "Pharmacy as a Clinical Profession." Hepler advanced the idea of clinical pharmacists as drug experts who render drug therapy to patients, ensuring the safe and appropriate use of drugs. The conference's goal was to move clinical pharmacy from the margins of pharmacy into the mainstream flow of pharmacy practice. This conference ignited the spark that led to other conferences.[23]

At the AACP annual conference at Charleston, South Carolina in 1987, Hepler, building on the work of Donald Brodie, introduced "pharmaceutical care" as the concept that guided clinical pharmacy. In 1989, at the second "Pharmacy in the 21st Century Conference" held in Williamsburg, Virginia,

Hepler teamed with Linda Strand to define pharmaceutical care as "the responsible provision of drug therapy for the purpose of achieving definite outcomes that improve a patient's quality of life."[24] Since that time pharmaceutical care has come to mean different things to different constituencies in pharmacy. Similar to the concept of clinical pharmacy, with pharmaceutical care, it is easier to define what it is not than what it is. Since the publication of Hepler and Strand's seminal paper on pharmaceutical care the profession has grappled with its implications as a philosophy, technique, and model of practice.[25]

Medication Therapy Management

Inspired by the concept of pharmaceutical care in the 1990s the scope of clinical pharmacy expanded, especially in hospital settings. When Congress passed the Medicare Prescription Drug, Improvement and Modernization Act in 2003 for senior citizens, the concept of medication therapy management (MTM) was born. With 11 pharmacy organizations working together the following definition of MTM was developed. "Medication therapy management is a distinct service or group of services that optimize therapeutic outcomes for individual patients. Medication therapy management services are independent of, but can occur in conjunction with, the provision of medication product."[26] By 2007, the American Medical Association included codes that would compensate pharmacists for cognitive services (medication advice) in its *Current Procedural Terminology*, which created a mechanism for MTM to become a reality.

Chapter Summary

Clinical pharmacy emerged out of the growth of the modern American hospital in the 1920s. Hospital pharmacists, who were a minority in their own profession, often were marginalized by their more numerous community pharmacy colleagues. Led by visionaries and activists, including Harvey A.K. Whitney, Donald Francke, Donald Brodie, and others, hospital pharmacists pioneered the idea of clinical pharmacy that would come to dominate the pharmacy profession by the end of the twentieth century. From the clinical pharmacy movement, the concepts of pharmaceutical care and MTM are now the guiding forces of twenty-first-century pharmacy practice.

Key Terms

E.C. Austin	Edward Swallow	formulary
John C. Krantz, Jr.	Harvey A.K. Whitney	Louis C. Zopf

Edward Spease	Eugene V. White	C. Douglas Hepler
American Society of Hospital Pharmacists (ASHP)	clinical pharmacy	Linda Strand
	Mary Anne Koda-Kimble	pharmaceutical care
Donald Francke	Joseph DiPiro	medication therapy management (MTM)
Gloria Neimeyer	American College of Clinical Pharmacy (ACCP)	
Donald Brodie		
David Burkholder		
The Ninth Floor Project	Hilton Head Conference	

Chapter in Review

1   Explain how the rise of the modern hospital in the 1920s gave rise to clinical pharmacy.

2   Explain why it took so long for clinical pharmacy to become accepted in the pharmacy profession.

3   Identify and explain the roles played by the pioneers of hospital pharmacy in the United States and how they became advocates for clinical pharmacy.

4   Describe the Ninth Floor Project at Moffit Hospital and its role in defining the practice of clinical pharmacy.

5   Explain what pharmaceutical care is and how it was introduced and received in the pharmacy profession.

6   Define what medication therapy management is and how it came about.

Notes

1 Gregory Higby, "From Compounding to Caring: An Abridged History of American Pharmacy." In *Pharmaceutical Care*, edited by Calvin H. Knowlton and Richard P. Penna (Bethesda, MD: ASHP, 2003), 23.

2 Charles E. Rosenberg, *The Care of Strangers: The Rise of America's Hospital System* (Baltimore, MD: Johns Hopkins University Press, 1987), 5–7.

3 George Griffenhagen, *150 Years of Caring: A Pictorial History of the American Pharmaceutical Association* (Washington, DC: APhA, 2002), 55.

4 Glenn Sonnedecker, comp., *Kremers and Urdang's History of Pharmacy* (Madison, WI: AIHP, 1976), 208.

5 Robert M. Elenbaas and Dennis B. Worthen, eds., *Clinical Pharmacy in the United States: Transformation of a Profession* (Lenexa, KS: American College of Clinical Pharmacy, 2009), 14.

6 George Griffenhagen, *150 Years of Caring*, 55.

7 Dennis B. Worthen, *Heroes of Pharmacy: Professional Leadership in Times of Change* (Washington, DC: APhA, 2008), 226–227.

8 Robert M. Elenbaas and Dennis B. Worthen, "Transformation of a Profession: An Overview of the 20th Century." In *Pharmacy in History*, edited by Gregory Higby (Madison, WI: AIHP),Vol. 51 (2009), No. 4, 152–153.

9 George Griffenhagen, *150 Years of Caring*, 56.

10 Robert M. Elenbaas and Dennis B.Worthen, eds., *Clinical Pharmacy in the United States*, 16.

11 Dennis B.Worthen, *Heroes of Pharmacy*, 227.

12 Ibid., 106–107.

13 Robert M. Elenbaas and Dennis B.Worthen, eds., *Clinical Pharmacy in the United States*, 21.

14 Robert M. Elenbaas and Dennis B.Worthen, eds., *Clinical Pharmacy in the United States*, 20.

15 Dennis B.Worthen, *Heroes of Pharmacy*, 108.

16 Ibid., 109.

17 Robert M. Elenbaas and Dennis B.Worthen, eds., *Clinical Pharmacy in the United States*, 31.

18 Gregory Higby,"From Compounding to Caring," 39.

19 Robert A. Buerki and L.V. Vottero, *Ethical Responsibility in Pharmacy* (Madison, WI: AIHP, 1994).

20 Robert M. Elenbaas and Dennis B.Worthen, eds., *Clinical Pharmacy in the United States*, 46.

21 Ibid., 75.

22 Robert M. Elenbaas and Dennis B.Worthen, "Transformation of a Profession," 173.

23 Ibid., 163–164.

24 Charles D. Hepler, "Reflections on Clinical Pharmacy: The Contribution of Pharmaceutical Care to Clinical Pharmacy." In *Clinical Pharmacy in the United States*, edited by Robert M. Elenbaas and Dennis B.Worthen, 148.

25 Jeanine K. Mount, "Pharmacy's Social Movement at the Turn of the Century: Introduction to Pharmaceutical Care Symposium." In *Pharmacy in History*, edited by Gregory Higby,Vol. 43 (2001), Nos. 2 & 3, 68.

26 B.M. Bluml, "Definition of Medication Therapy Management: Development of Professionwide Consensus," *Journal of the American Pharmacists Association* (Washington, DC: APhA) Vol. 45 (2005), 566–572.

Index

H + S
Embl
HOOPS DATES
Book Photo
OFF
Im SLU

PACK

Gifts
LUETIN
DVD X mas part
Mout - dinner
VSA But ale
Dhan Site PC
Library Cont & DC
Farm Address Vntn
Chris Eng Exam
Prnest . Gift